ENVIRONMENTAL SCIENCE, ENGINEERING AND TECHNOLOGY

NUTRIENTS, WASTEWATER AND LEACHATE

TESTING, RISKS AND HAZARDS

Environmental Science, Engineering and Technology

Additional books in this series can be found on Nova's website under the Series tab.

Additional e-books in this series can be found on Nova's website under the eBooks tab.

ENVIRONMENTAL SCIENCE, ENGINEERING AND TECHNOLOGY

NUTRIENTS, WASTEWATER AND LEACHATE

TESTING, RISKS AND HAZARDS

AMIMUL AHSAN
EDITOR

nova
science publishers
New York

Library of Congress Cataloging-in-Publication Data

ISBN: 978-1-53613-949-5

Published by Nova Science Publishers, Inc. † New York

CONTENTS

PREFACE

The theoretical analysis and modeling on nutrients in water catchments and in irrigation water is a significant issue in a design of wide range of industrial processes and devices. Besides, the amount of antibiotic resistant fecal coliform bacteria in wastewater treatment plants is important to monitor and control. In addition, the hazardous landfill leachate needs to treat and handle properly to save our environment and public health. These three ideas are presented in this book.

This book introduces advanced ideas on *Nutrients, Wastewater and Leachate: Testing, Risks and Hazards* to the international community. It includes six advanced and revised contributions, and it covers (1) nutrients and sediment loads in water catchments with heterogeneous land uses, (2) effect of nitrogen release from bioplastic fertilizer (BpF) composites: slow release fertilizer control application, (3) prevalence of antibiotic resistant fecal coliform bacteria in wastewater treatment plants, (4) antibiotic resistant coliforms - from human gut to wastewater, (5) landfill leachate treatment by electrolysis and preparation of low cost activated carbon, and (6) treatment of scheduled waste landfill leachate: perspective in Malaysia.

The readers of this book will appreciate the current issues on water vapor condensation modeling in different aspects. The approaches would be applicable in various industrial purposes as well. The advanced idea and information described here will be fruitful for the readers to find a sustainable solution in an industrialized society.

The editor of this book would like to express sincere thanks to all authors for their high quality contributions and in particular to the reviewers and assistants for reviewing/revising the chapters.

Amimul Ahsan
Dhaka, Bangladesh
January 31st, 2018

Acknowledgments

All praise be to Almighty Allah, the Creator and the Sustainer of the world, the Most Beneficent, Most Benevolent, Most Merciful, and Master of the Day of Judgment. He is Omnipresent and Omnipotent. He is the King of all kings of the world. In His hand is all good. Certainly, over all things Allah hast power.

The editor would like to express appreciation to all who have helped to prepare this book. The editor expresses the gratefulness to Carra Feagaiga, Stephanie Gonzalez, Richard Schortemeyer III, Frank Columbus and Nadya Gotsiridze-Columbus, Nova Science Publishers, Inc., NY, USA. In addition, the editor appreciatively remembers the assistance of all authors and reviewers of this book.

Gratitude is expressed to Mrs. Ahsan, Ibrahim Bin Ahsan, Marium Binti Ahsan, Yahya Bin Ahsan, Yusuf Bin Ahsan, Mother, Father, Mother-in-Law, Father-in-Law, Brothers and Sisters for their endless inspirations, mental supports and also necessary help whenever any difficulty.

Amimul Ahsan
Dhaka, Bangladesh
January 31st, 2018

In: Nutrients, Wastewater and Leachate
Editor: Amimul Ahsan
ISBN: 978-1-53613-949-5

Chapter 1

NUTRIENTS AND SEDIMENT LOADS IN WATER CATCHMENTS WITH HETEROGENEOUS LAND USES

S. K. Das[1,*], A. W. M. Ng[1], B. J. C. Perera[1] and A. Ahsan[2,3]
[1]College of Engineering and Science,
Victoria University, Melbourne, Australia
[2]Department of Civil Engineering,
Uttara University, Dhaka, Bangladesh
[3]Department of Civil and Construction Engineering,
Swinburne University of Technology, Melbourne, Australia

ABSTRACT

Understanding how nutrient and sediment concentrations in leachates (e.g., agricultural runoff) vary with land uses is critical to study the

---

[*] Corresponding Author Email: sushil.das@live.vu.edu.au.

current and future impacts of land use changes on catchment water quality. This chapter presents a simple procedure on how to use specific flow regimes to correlate pollutant load with land use activities in heterogeneous catchments, even in shortage of specific event-based water quality grab sample data. The procedure uses a simple baseflow separation technique and a data-based model with other readily available data and tools. The specific tasks of this study were: (1) establishment of whether there is any statistically significant difference between pollutant concentrations in water quality samples representing baseflow conditions and runoff events, (2) estimation of nutrient and sediment loads, and (3) identification of significant and major sources of pollutant loads. The proposed procedure was then applied to the Yarra River catchment in Victoria, Australia, and found to be effective, particularly in circumstances where limited time (for complex analysis) and data are available. The study found that at all monitoring stations, pollutant concentrations were significantly ($p \leq 0.01$) greater in runoff events than in baseflow conditions, and the LOADEST regression models developed explained more than 85% variability in loadings. Overall, water quality and pollutant concentrations were influenced by specific flow regimes and land uses. Major pollutant loads were generated from agricultural and urban runoff, i.e., non-point sources. About 65%, 69% and 86% of annual TN, TP and TSS total loads respectively at the outlet of the catchment were delivered by runoff events where runoff event days were on average 18% and 31% only at the main stream and tributary stations respectively during the study period.

Keywords: specific flow regimes, Pollutant concentrations, pollutant loads, non-point source, LOADEST, SWAT

1. Introduction

Scarcity of water, deterioration of water quality and excessive sediments in rivers and creeks have become challenging issues for food supply, food security, human health and natural ecosystems. This is particularly the case with rapid changes in land use and farming practices. In the last few decades, changes in land use patterns caused by demographic, economic, political and/or cultural mutations have notable effects on water supply, water quality and soil erosion [1]. A consequence of the conversion of tropical rainforests to pastures or cultivated land results in a decrease of the porosity of the top-soil where organic matter and nutrients concentrate. This leads to more runoff, nutrient leaching and erosion, responsible for an important reduction of on-site fertility and off-site consequences. Some examples of such off-site consequences are water pollution by nutrients which in turn increases the eutrophication hazards, flooding, decrease of groundwater recharge, and fill of valley bottoms or reservoirs. All these on-site and off-site effects may dramatically jeopardize the future of natural ecosystems and the economic development of society.

Eutrophication is a continuous problem for water resource managers as it leads to significant loss of amenity in terms of water supply, fisheries and recreation [2]. The major nutrients that cause eutrophication are phosphorus and nitrogen [3]. Phosphorus is more often limiting in freshwater systems, whereas nitrogen is more important in marine systems [4]. In the past, nutrients originated from point source effluents have been the main contributors. Point sources are relatively easy to identify, quantify and control through advanced treatment and legislative actions. However, even though point source pollution is under control, the limnological problem has not diminished due to increases in population, recent land use development and inappropriate application of farming chemicals. Attention has now switched to non-point nutrient sources originated from intensive agriculture and urban runoff. In contrast to a point source, where a known volume of contaminant is discharged from a single identifiable source such as a sewage treatment plant, non-point source pollution is an aggregate of small contaminant inputs distributed within a catchment through rural runoff, urban stormwater runoff, or failed septic systems.

Catchment-scale management programs have been proven to be efficient in reducing water pollution from land use activities [5]. The water catchments are generally heterogeneous with respect to their land uses. Understanding the connection between different land uses, land management, and water quality is important in the development of appropriate management strategies of a catchment. However, lack of information of nutrient and sediment loads, their sources and correlations with land uses at various sections of a catchment are major impediments in applying these management programs. Complex mechanistic models such as SWAT [6], HSPF [7] are generally used to quantify the generation and movement of these pollutants with the specific mechanisms causing surface runoff and pollutant transport. These models are also used to simulate alternative best management practices for future land use changes. However, these models require extensive data inputs, high computational power and expertise. Alternatively, simple data-based models such as the regression model LOADEST [8]; export coefficient model [9] or generation rates-based model EMSS [10] can be used to estimate pollutant mass loads and their relationships with land uses, especially for management use where limited time (for complex analysis) and limited data available [11]. The data-based models are based on the analysis of observations, and seek to characterize response from these data [12]. The computational and data requirements for such models are usually less than for mechanistic models, and are often supported by coarse measurements. However, the data-based models cannot simulate alternative best management practices for future climate conditions and land use changes.

Pollutant export coefficients or generation rated-based approaches are most commonly used approaches to estimate pollutant loads relating specific land use types. The use of pollutant export coefficients for estimating pollutant loads is based on the knowledge that, for a given climatological regime, specific land use types will yield or export characteristic quantities of pollutants to a downstream water body over an annual cycle or longer term. Typically, pollutant export coefficients are determined by monitoring pollutant exports from small catchments with a predominant land use or by using field plots to isolate individual land uses [13]. However, most catchments (even small ones) are generally comprised of a variety of different land uses, i.e., heterogeneous. Therefore, these export coefficients do not represent the average of conditions and practices (e.g., soils, slopes, tillage practices and proximity to streams) within the entire catchment. Multiple regression techniques [14] can be used to develop pollutant export coefficients for the major land uses in a heterogeneous catchment, but it requires 30 or more water quality stations to provide adequate power to the regression analysis [15]. In general, pollutant export coefficient approach requires large number of water quality monitoring stations representing specific land use type. This necessitates greater effort in determining pollutant export coefficients and may result in larger uncertainties in load estimation.

The generation rates-based approach can also be used to estimate loads using event mean concentration (EMC) for stormflow and dry weather concentration (DWC) for baseflow as used in the simple model EMSS [10]. The model estimates daily runoff, which is partitioned into event and baseflow components. These flow components are multiplied by user specified generation rates (EMC and DWC) to estimate daily loads. This approach requires event-based water quality data collection which is very sparse. In most cases, water quality measurements are done sparsely, mainly for compliance purposes. Moreover, because of spatial heterogeneity of a catchment, different EMC and DWC values need be allocated to sub-catchments depending on land use type, requiring large number of water quality monitoring stations to estimate different EMC and DWC values. Further research, supported by event-based water quality data collection, is required to refine this approach [16].

Specific flow regimes (i.e., baseflow conditions and runoff events) defined by a simple hydrograph separation technique can provide an insight into pollutant sources, transport, and potential impact on a river [17] in a simple way with existing monitoring stations and collected water quality grab sample data by categorizing water quality grab sample data and estimated pollutant loads into the specific flow regimes. For example, stream water quality during baseflow conditions will be more affected by discrete pollution sources such as municipal wastewater treatment plants, non-point pollution sources from groundwater recharge, and various abiotic and biotic processes within the stream system. Alternatively, water quality during surface runoff events will be more influenced by resuspension of temporarily retained in-stream constituents and non-point source pollutants entrained by surface runoff during episodic precipitation events. This procedure can define specific flow regimes even in the absence of specified event-based water quality grab sample data. Hydrograph separation allows the separation of water quality data and estimated loads into that collected during baseflow conditions and runoff events. For this purpose, a simple baseflow separation technique is used to define each day of the study period to a specific flow regime using the following conditions; (i) baseflow conditions are defined as those days when baseflow contributes ≥70% of daily streamflow, and (ii) runoff events are defined as those days when baseflow contributes <70% of daily streamflow [17, 18]. The word "runoff" for "runoff events" and "baseflow" for "baseflow conditions" will be used interchangeably throughout the chapter. Accordingly, water quality grab sample data collected on a baseflow day is tagged as baseflow data, and if collected on a runoff day is tagged as runoff data at each station. Similarly estimated daily pollutant loads from the data-based model such as LOADEST can be categorized into baseflow loads those fall on the baseflow days and runoff loads those fall on the runoff days in the study period at each station.

This chapter presents a simple procedure on how to use specific flow regimes to correlate pollutant load with land use activities in a catchment. The procedure used a data-based model LOADEST to estimate pollutant loads and a simple baseflow separation technique to define specific flow regimes with other readily available data and tools. The specific tasks of this study were: (1) to establish whether there is any statistically significant difference between pollutant concentrations in water quality grab samples (Total nitrogen-TN, Total phosphorus-TP and Total suspended solids-TSS) representing baseflow conditions and runoff events, (2) to estimate pollutant loads: TN, TP and TSS at various sections of a catchment, and (3) to identify the significant and major sources of the pollutant loads. This procedure was then applied to the Yarra River catchment in Victoria, Australia and demonstrated as a case study. Two simple baseflow separation software tools Base Flow Index (BFI) [19] and Baseflow Filter Program (BFP) [20] were evaluated comparing them with a complex model SWAT to investigate the efficiency of a simple baseflow separation technique in the case study. The outcome of this case study can be used in the initial planning stage of developing appropriate management strategies for the Yarra River catchment.

The methods and techniques used in the proposed procedure are first discussed in this chapter, describing the selection of water quality monitoring stations and catchment subdivision, the identification of major land uses in sub-catchments, the analysis of specific flow regimes, and the load estimation methods. The case study of the Yarra River catchment in Victoria which demonstrated the proposed procedure in assessing water quality is described then, followed by the results. Finally, the conclusions of the study are presented.

2. Methods and Techniques

2.1. Selection of Water Quality Monitoring Stations and Catchment Subdivision

For proper evaluation of correlation between mass loads of pollutants and land use in a catchment, the water quality monitoring stations should be selected in such a way that these represent the whole catchment uniformly. Sampling stations located in the main river and the major tributaries are useful to enable division of the whole catchment uniformly into sub-catchments. This helps to identify critical sub-catchments responsible for yielding high pollutant loads. In case of multiple sampling stations close to each other in any particular tributary, the best sampling station can be selected based on measured data characteristics (censored or missing) and periods of record.

Once the water quality monitoring stations are selected, the catchment can be divided into sub-catchments using the Digital Elevation Model (DEM) data and the Geographic Information System (GIS) software based on the monitoring stations. Each sub-catchment defines the drainage area above each monitoring station. Relatively high resolution and good quality global scale DEMs have become available recently in public domain. The CGIAR-CSI GeoPortal is able to provide SRTM 90m Digital Elevation Data for the entire world (http:// srtm.csi.cgiar.org/). ASTER 30m Global Digital Elevation Model (GDEM), jointly developed by The Ministry of

Economy, Trade, and Industry (METI) of Japan and the United States National Aeronautics and Space Administration (NASA), can be downloaded from NASA's Earth Observing System (EOS) data archive (http://asterweb.jpl.nasa.gov/gdem-wist.asp). The choice of the DEM resolution depends on the system availability (or the computing capacity) and the accuracy needed to locate the water quality sampling stations in order to delineate sub-catchments.

2.2. Identification of Major Land Uses in Sub-Catchments

The water catchments are generally heterogeneous with respect to their land use types. It is widely recognized that land uses have significant impact on surface water quality. Different sub-catchments have different land uses in general. Therefore, it is essential to identify the major land use types in each sub-catchment. Forest, agriculture, and urban are the predominant land use types which significantly affect the pollutant loadings. The proportion of major land uses in a sub-catchment above a monitoring station can be determined by overlaying a land use map into the drainage network (sub-catchment) of that monitoring station using spatial tools viz GIS. Land cover/land use maps can often be obtained from government agencies or can be downloaded free of charge. The European Commission Joint Research Centre has provided 45 classes of global land cover data for year 2000 (http://forobs.jrc.ec.europa.eu/products/glc2000/glc2000.php). If possible, locally available land use information should be used.

2.3. Analysis of Specific Flow Regimes

Hydrograph separation allows the separation of water quality data and estimated loads into that collected during baseflow conditions and runoff events. For this purpose, a simple baseflow separation technique is used to define each day of the study period to a specific flow regime using the following conditions; (i) baseflow conditions are defined as those days when baseflow contributes ≥70% of daily streamflow, and (ii) runoff events are defined as those days when baseflow contributes <70% of daily streamflow [17, 18].

Various techniques are available to separate baseflow from gauged streamflow data. These include traditional manual graphical procedures to more recent automated procedures [21]. Graphical separation methods tend to focus on defining the points where baseflow intersects the rising and falling limbs of the quickflow response. Manual separation of the streamflow hydrograph into surface flow and baseflow is difficult and inexact; often results derived from such manual methods cannot be replicated by different investigators [22]. On the other hand, automated data processing or filtering procedure has no physical meaning, but it removes the subjective aspect from manual baseflow separation, and it is fast, consistent, and reproducible [23]. For an example, the computer software program Base Flow Index (BFI) [19] estimates the baseflow using an automated technique (known as smoothed minima filtering technique) developed by the Institute of Hydrology [24]. BFI uses the minimum daily streamflow in five consecutive days, and minimum flows less than 90% of adjacent minimum flows are defined as turning points. The turning points are used to interpolate the baseflow hydrograph [17]. The automated procedures predominantly involve the use of another filtering procedure known as recursive digital filters that have their basis in signal analysis and processing, and are used to remove the high-frequency quickflow signal to derive the low-frequency baseflow signal [25]. Baseflow Filter Program (BFP) is an automated digital filter based software developed by USDA-AES [20]. Several studies have shown that the automated digital filter technique compares well with manual and other automatic baseflow separation techniques and with measured results [23, 25, 26].

Integrated groundwater and surface water physically based process models offer an effective approach to estimate baseflow. The explicit modeling of physical processes in models such as SWAT [6] can provide one of the best means of quantifying the absolute volume of groundwater flow. However, these approaches are very demanding in terms of both resources and data due to the complexity of the models, and do not provide a simple method that can be rapidly applied across numerous case study locations.

After categorising water quality grab sample data of the whole study period into baseflow or runoff data, summary statistics including maximum, minimum, median and mean values are used to describe the two sampling data sets for each pollutant at each station. Then statistical methods are used to identify whether there is any statistically significant difference between concentrations in baseflow and runoff water quality data for each pollutant. The Mann-Whitney rank-sum test, a nonparametric procedure and the t-test are generally used for determining whether two independent data groups differ significantly or not [27]. SPSS statistical analysis software [28] can be used to do these tests. The t-test is perhaps the most widely used method for comparing two independent groups of data. It is familiar to most water resources scientists. However, there are five often overlooked problems with the t-test that make it less applicable for general use than the nonparametric Mann-Whitney rank-sum test [27]. These are: 1) lack of power when applied to non-normal data, 2) dependence on an additive model, 3) lack of applicability for censored data, 4) assumption that the mean is a good measure of central tendency for skewed data, and 5) difficulty in detecting non-normality and inequality of variance for the small sample sizes common to water resources data. The tests dealing with more than two groups of independent data (similar to the above tests) are Kruskal-Wallis test [27] and analysis of variance (ANOVA) respectively, where ANOVA has the same limited applicability as discussed above.

2.4. Estimation of Pollutant Loadings

The load of a water quality pollutant in a stream (i.e., the weight of material transported during specific time period) is a function of the concentration of the pollutant and the stream discharge. In general, load L (kg), over a time period T, can be represented by the equation:

$$L = \int_0^T \mathrm{CQ}dt \tag{1}$$

where C (mg/l) is the pollutant's concentration and Q (m^3/s) is the water discharge. The use of automated equipment allows precise discharge measurements economically for short time intervals (e.g., hourly or less), but water quality constituents are generally collected as grab samples at regular intervals (i.e., weekly, monthly or seasonally). The measurement of a pollutant concentration requires water sampling, storing, and costly laboratory analyses, which makes concentration measurement the limiting factor for estimating pollutant loads. Therefore, in most cases, water quality measurements are done sparsely, mainly for compliance purposes. In this case, integrating concentration and streamflow may be inappropriate to calculate the loads. There are many different techniques used for calculating load estimates, differing in complexity, accuracy and bias. The choice of technique depends on the data resolution, the mathematical ability of the operator, the computer technology available, and the relationship between concentrations of various pollutants and streamflow data [29]. Existing data-based methods for load estimation using field data can be classified into three major classes: (i) averaging estimators, (ii) ratio estimators, and (iii) regression methods [30]. The averaging estimators, also called integration or interpolation methods, are based on some form of averaging concentration or flow data. The ratio estimators are based on the ratio of flow and concentration and often modified by a bias correction factor. The regression methods are based on fitting a relationship between flow and concentration for estimating a continuous trace of concentration. Reviews of methods of load estimation techniques using field data can be found in Letcher et al. [29]; Mukhopadhyay and Smith [31]; Quilbe et al. [32]; and Marsh and Waters [30].

Based on the literature, Quilbe et al. [32] suggested that: (i) averaging estimator methods are accurate only when concentration measurements are available for the entire flow range; (ii) ratio estimator methods are less sensitive to river and pollutant characteristics than regression methods but requires more data to achieve the same level of precision, and they are robust and unbiased under systematic sampling, as well as under stratified event sampling; (iii) regression methods can give the best results for sediment and total phosphorus if streamflow and concentration data are strongly correlated for a wide range of streamflow values. The regression methods do not require an extensive database but the quality of prediction depends on the quality of the correlation between flows and concentrations [33]. Therefore, the regression method should be given priority if the correlation is high enough. Since the temporal variability of the relationship between concentration and streamflow can be important, some authors [34] define regression equations as functions of time in order to take into account nonlinearities as well as seasonal and long-term variability.

3. Case Study

The Yarra River catchment in Victoria (Australia) is used as the case study in this chapter.

3.1. Study Site Description

The Yarra River is one of the major rivers in Victoria, Australia, and is considered as one of the valuable assets for all Victorians [35]. The Yarra River catchment, as shown in Figure 1 has an area of over 4,000 square kilometres and the distance from the head at the Yarra Ranges National Park, to the end of its estuary at Port Phillip Bay, is about 245 km, with additional 1,800 km of named tributaries [35]. About 21 percent of the catchment retains its natural vegetation, 57 percent is agricultural and 22 percent is urbanised [36].

The Yarra River catchment is an important water resources catchment for Victoria, where over one-third of Victoria's population (approximately 2 million) lives in this catchment. The catchment water resources support a range of uses valued by the Melbourne's community, including urban water supply, agricultural and horticultural industries, and downstream user requirements as well as flow requirements for maintaining environmental flows. There are seven major reservoirs in the catchment. These reservoirs are used mainly for urban water supply and storage purposes.

Three distinct segments, namely Upper, Middle and Lower Yarra, have been defined for the Yarra River catchment, based on different land use activities, as shown in Figure 1. The Upper Yarra segment, from the Yarra Ranges National Park to the Warburton Gorge at Millgrove, consists of mainly dense and extensive forested area with minimum human population. Devonian sedimentary rocks with coarse bed material mainly dominate the area [37]. This river segment has the best water quality since the area is protected for the sole purpose of harvesting water for human consumption. The Middle Yarra segment, from the Warburton Gorge to Warrandyte Gorge, consists of mainly rural floodplains and valleys with limited urban development. A mixture of Devonian granite rocks, Silurian and Devonian sedimentary rocks and Tertiary basalts dominate this part of the catchment area. The bed material throughout this segment consists of fine sediments [37]. The majority of the surrounding land in this river segment is used for agricultural purposes, causing major non-point source pollution in terms of high nutrient runoff [36]. The Lower Yarra segment, downstream of Warrandyte, consists of mainly urbanised floodplains, and has the poorest water quality because of urban runoff. The geology in this segment includes Devonian plutonic rocks, Palaeozoic sedimentary rocks, Newer volcanic and Quaternary sediments. Both sand and compacted clay are found in the river bed.

The Yarra River catchment has a temperate climate. The annual rainfall of the Yarra River catchment varies from approximately 1,600 mm in the Upper Yarra area to about 600 mm in the Lower Yarra region [38]. Analysis of Yarra River flows by Ng and Perera [39] showed that low flows occur from November to June, whereas high flows occur during the other times of the year. A major diversion of approximately 51.3 GL/year (on average) occurs at Yering Gorge [37]. Due to increases in population, recent land use development in the catchment and inappropriate application of farming chemicals, the river water quality had degraded, and water quality at most monitoring stations does not comply with state environment protection policy (SEPP) objectives in the last few years [40]. This degradation of water quality in the Yarra River has prompted a need to assess nutrients and sediment loads at different sections of the catchment for development of appropriate management strategies to improve water quality.

Three pollutants namely Total Nitrogen (TN), Total Phosphorus (TP) and Total Suspended Solids (TSS) were considered in this case study. In the Yarra River catchment, phosphorus is the key pollutant in the waterways whereas in Port Phillip Bay, nitrogen is the key nutrient affecting algal growth and must be managed to maintain the health of the bay [36, 40, 41]. Based on the CSIRO Port Phillip Bay Environmental Study [41] and SEEP objectives, Melbourne Water and EPA, Victoria set different short-term and long-term target for rivers and creeks in the Port Phillip Bay [40]. These include reduction of nitrogen loads to Port Phillip Bay at least 40 tons per year from diffuse catchment sources (offset urban growth) by 2014; by 2015 the Environmental Water Reserve for the Yarra system will be increased by 17GL; and by 2025, all natural rivers and creeks to be in good or better condition.

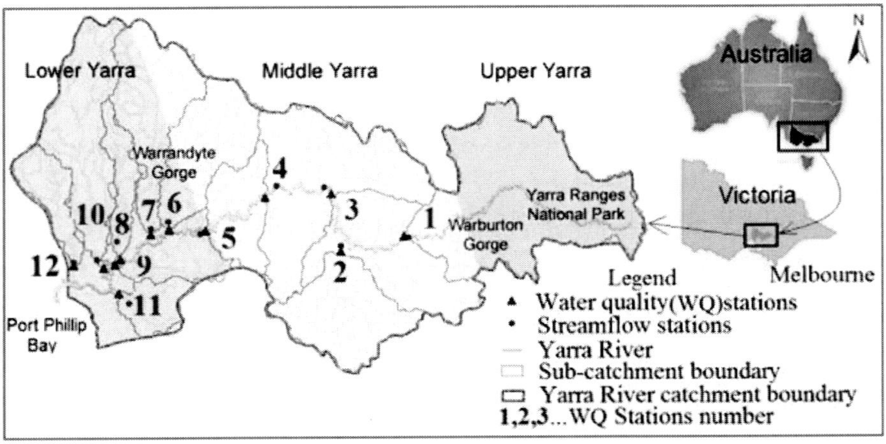

Figure 1. Yarra River catchment with the monitoring stations.

3.2. Data Collection and Preliminary Analysis

The major water quality monitoring program in place in the Yarra River catchment is the Melbourne Water Corporation's monitoring network (incorporating the EPA Victoria Fixed Site Network). Under this program, water quality is monitored monthly at 33 stations along the Yarra River and its tributaries. Of the 33 stations, 21 stations have data from 1994 where as other 7 stations have data from 1995, and the remaining 5 stations have data from 1998. Of these 33 stations, 12 stations were considered in this study in such a way that these stations represent the whole catchment uniformly. Among these 12 stations, 5 stations were on the main stream of the Yarra River, and the remaining stations were on the tributaries as shown in Figure 1. Data characteristics (whether any pollutant data is censored or missing) and period of records available also affected the selection of the stations. Based on the above criteria, the study period was chosen from January 1998 to June 2009 (except June 1999 to June 2009 for station 8). Water quality data on concentrations of TN, TP and TSS obtained through grab samples were collected for these 12 stations from Melbourne Water Corporation. The corresponding daily streamflow data for these 12 stations also were collected from the nearby 12 streamflow stations of Melbourne Water Corporation.

Table 1 shows the water quality monitoring stations with their position and their nearby streamflow gauging stations. The first five stations (stations 1, 3, 4, 5, and 9) as shown on Table 1 are on the Yarra River main stream, and others are on the tributaries. It also compares the daily streamflow between the study period and the whole period of record for each gauging station. In general, the study period is very dry compared to the whole period of record which indicates runoff contribution into the total streamflow is comparatively less in the study period. Table 1 also shows that the main stream stations are drier than the tributary stations. Because of this, on average the percentage of runoff days in the main stream stations is lower than in the tributary stations. This can be seen in Table 2 where the average percentages of runoff days in the main stream and tributary stations are 18 and 31 respectively. Moreover, since a significant amount of water is pumped out at Yering Gorge Pumping Station above station 5 [37] as discussed in Section 3.1, the daily steamflow of this station is lower than the upstream station 4 (Table 1).

Table 1. Water quality stations and streamflow statistics (Station number corresponds to Figure 1)

Station Number		Water quality stations	Easting	Northing	Period of streamflow record	Daily streamflow (m^3/s)			
						Study period 1998-2009			Ratio*
						Min.	Mean	Max.	
Main stream	1	Yarra River at Millgrove	380712	5820934	1970-2009	0.709	2.87	37.14	1.57
	3	Yarra River at Healesville	367012	5828984	1980-2009	0.962	6.44	79.40	1.66
	4	Yarra River at Coldstream	354512	5828284	1991-2009	1.346	7.89	134.59	1.47
	5	Yarra River at Warrandyte	343212	5821984	1970-2009	1.285	6.70	113.28	1.92
	9	Yarra River at Kew	326212	5815784	1975-2009	1.547	8.52	218.57	1.62
Tributary	2	Woori Yallock Creek at Woori Yallock	368752	5818084	1975-2009	0.084	1.44	34.28	1.74
	6	Diamond Creek at Eltham	336412	5822424	1978-2009	0.001	0.37	91.98	1.72
	7	Plenty River at View Bank	333012	5821384	1977-2009	0.001	0.30	46.74	2.00
	8	Darebin Creek at Alphington	327269	5816798	1999-2009	0.001	0.35	50.33	0.97
	10	Merri Creek at Yarra Bend	324192	5815224	1975-2009	0.040	0.72	106.13	1.56
	11	Gardiners Creek at Hawthorn	326912	5810384	1978-2009	0.001	0.56	48.28	1.35
	12	Moonee Ponds Creek at Flemington	318412	5815684	1991-2009	0.013	0.42	87.17	1.02

* Ratio = (Record period mean)/(Study period mean).

Table 2. Percentage of major land uses, and baseflow condition and runoff event days designated by BFP

Station	Number	Period of Study	Land uses associated with the stations					Total Days in Study Period		% of Runoff days	
			Forest (%)	Agri-culture (%)	Urban (%)	Others (%)	Drainage area (km^2)	Baseflow condition days	Runoff event days		
Main stream	1	1998-2009	64	35	0	1	844	3560	639	15	Average = 18
	3	1998-2009	50	44	6	0	1566	3450	749	18	
	4	1998-2009	49	44	7	0	2153	3515	684	16	
	5	1998-2009	46	43	10	1	2354	3412	787	19	
	9	1998-2009	38	43	18	1	3323	3262	937	22	
Tributary	2	1998-2009	33	49	18	0	326	3462	737	18	Average = 31
	6	1998-2009	26	49	25	0	309	2663	1536	37	
	7	1998-2009	24	53	21	2	351	2778	1411	34	
	8	1999-2009	3	52	44	1	105	2443	1240	34	
	10	1998-2009	2	67	31	0	392	2960	1239	30	
	11	1998-2009	0	0	100	0	107	2645	1554	37	
	12	1998-2009	9	36	49	6	125	3019	1180	28	

A sub-catchment map defining drainage area for each water quality monitoring station was generated from ASTER 30m global DEM (http://asterweb.jpl.nasa.gov/gdem-wist.asp) using ArcGIS 9.3 tools. A land cover map (50m grid raster data) collected from the Australian Bureau of Agricultural Resource Economics and Sciences (ABARES) was overlaid, and the proportions of major land uses in each sub-catchment were calculated. The size of the drainage area and percentage of major land uses above each monitoring station are presented in Table 2. In general, the dominant land use type in the tributary stations was either agriculture or urban, whereas it was forest-agriculture mix type in the main stream stations. Moreover, the percentage of agricultural and urban land uses increases from upstream to downstream of the catchment.

3.3. Analysis of Specific Flow Regimes

A SWAT model was developed for the mid-agricultural part of the Yarra River catchment, as shown in Figure 2. The SWAT model required DEM, land use, soil and climate data for model development, and streamflow data for model calibration. The spatial data were processed using the ArcSWAT tool (Figure 2). The climate data were collected from SILO climate database (http://www.longpaddock.qld.gov.au/silo/). All necessary spatial datasets and database input files for the model were organized and assembled following the guidelines of Winchell et al. [42]. ArcSWAT embedded sensitivity and auto-calibration tool was used for sensitivity analysis and calibration of the model. The outlet of the catchment at Warrandyte station (station 5 of Figure 1) was used for calibration. The SWAT model was calibrated (daily, monthly and annual R^2 values were 0.78, 0.93, and 0.96 respectively) and validated (daily, monthly and annual R^2 values were 0.74, 0.82, and 0.87 respectively) for total streamflow. At the same station, baseflow was separated from observed daily streamflow time series using the BFI and BFP software tools. The simple baseflow separation techniques compared well with the SWAT simulated baseflow. The R^2 between BFP generated and SWAT simulated baseflow were 0.73, 0.86 and 0.91 for daily, monthly and annual baseflow respectively. Similarly, R^2 between BFI generated and SWAT simulated baseflow were 0.57, 0.78 and 0.90 for daily, monthly and annual baseflow respectively.

Based on this comparison, the BFP software was then used to define specific flow regimes through baseflow separation. Figure 3 shows the hydrograph separation at Coldstream station (station 4) as a typical case for the years of 1998 and 1999 only. Each day of the study period at a station was designated to be either baseflow or runoff day as per the baseflow contribution criteria discussed in Section 2.3, and is shown in Table 2. The percentage of runoff days in the main stream stations was less than in the tributary stations (Table 2).

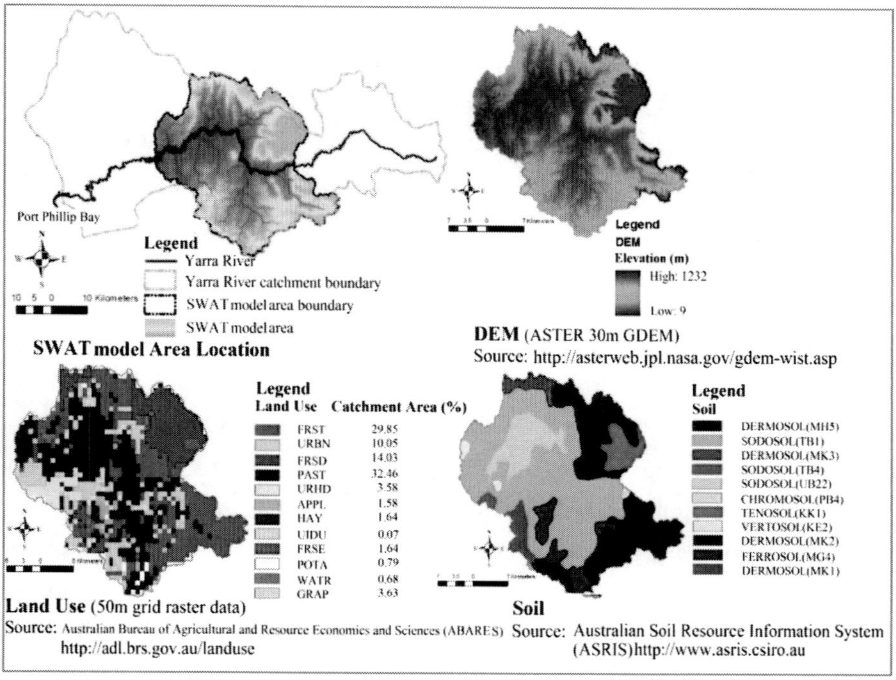

Figure 2. SWAT modeling area and spatial input data.

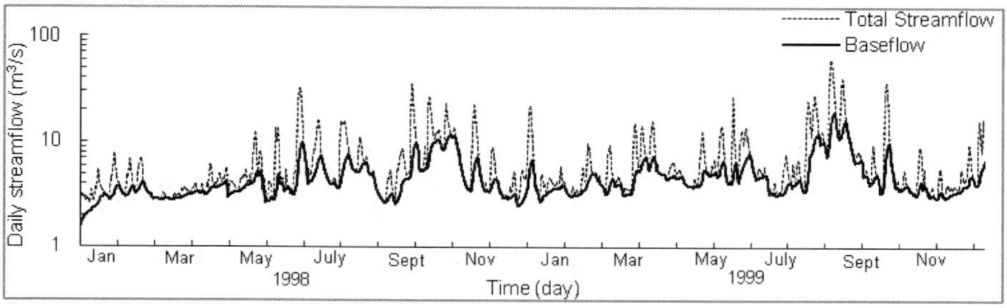

Figure 3. Streamflow hydrograph separation at Coldstream station (station 4).

After categorising water quality concentration data into those representing baseflow conditions or runoff events, summary statistics were calculated to describe the two sampling concentration data sets for each pollutant including maximum, minimum, median, and mean values. Then the Mann-Whitney rank-sum test was used to determine the statistically significant difference between baseflow and runoff pollutant concentrations at each monitoring station. Similarly, the Kruskal–Wallis test was used to determine the statistically significant difference among pollutant concentrations at different stations [17, 18]. For both tests, IBM SPSS statistical analysis software was used. These tests were selected because they do not require data to be normally distributed, and are applicable in case of censored data [27]. The medians of the independent data sets were described as being significantly different if the probability p-value of the test was less than or equal to 0.01.

3.4. Estimations of Pollutant Loads and Yields

The correlations between streamflow and concentrations of TN, TP and TSS were first determined for all water quality stations. On the main stream of the Yarra River, the correlation of TN was 0.6 to 0.8 and that of TP and TSS was 0.5 to 0.8, with the exception of TP at stations 3 and 9, and TSS at stations 3 and 4. Similarly, on the tributaries, the correlation of TN, TP and TSS was in the ranges of 0.4-0.7, 0.4-0.8, and 0.5-0.9 respectively with exception of TN at stations 11 and 12, and TP at stations 2 and 11. In general, all correlations were found to be significant ($p \leq 0.01$). The regression based model LOADEST [8] was then used to estimate pollutant loads. The LOADEST model is well documented, and is accepted as a valid means of calculating constituent load from a limited number of water quality measurements [43]. LOADEST considers regression equation as a function of time in order to take into account nonlinearities as well as seasonal and long-term variability. The regression model LOADEST has been widely used, particularly by the U.S. Geological Survey [17, 18, 44].

The U.S. Geological Survey developed the LOADEST, a FORTRAN program for estimating pollutant loads in streams and rivers, that accounts for many of the statistical challenges encountered when formulating, calibrating and applying regression models in estimation of pollutant loads [8]. The calibration and estimation procedures within LOADEST are based on three statistical estimation methods. The first two methods, adjusted maximum likelihood estimation (AMLE) and maximum likelihood estimation (MLE), are appropriate when the calibration model errors (residuals) are normally distributed. Of the two, the AMLE method is considered when the calibration data set (time series of streamflow, additional data variables, and concentrations) contains censored data. The third method, least absolute deviation (LAD), is an alternative to maximum likelihood estimation when the residuals are not normally distributed.

The LOADEST model evaluates the relationships between pollutant loads (dependent variables) and streamflow and time variables (explanatory variables) based on eleven predefined models and a user-defined model. The users can select a model manually or can automatically select the best model from the first nine models based on the lowest value of Akaike Information Criterion (AIC) [45]. In statistics, AIC is the most commonly used criterion for model selection from a set of parametric models.

The first nine models include different combinations of time variables and seasonal variables to consider time trend and seasonal trend in a continuous manner. For example, the seven-parameter model is given below;

$$\ln(L) = a_0 + a_1 \ln Q + a_2 \ln Q^2 + a_3 \sin(2\pi dtime) + a_4 \cos(2\pi dtime) + a_5 dtime + a_6 dtime^2 \tag{2}$$

where ln = natural logarithm; L = pollutant load in kg/day; a_0 = dimensionless regression constant; a_1, a_2, a_3, a_4, a_5, a_6 = dimensionless regression coefficients; Q = daily mean streamflow in cubic feet per second; and dtime = time parameter in decimal years from the start of the study period. Within the above model, the explanatory variables lnQ and $\ln Q^2$ account for the dependence on streamflow, the sine and cosine terms account for seasonal variability, and the dtime and $dtime^2$ account for the time trend. For abrupt seasonal change, users can use last two predefined models or develop a user-defined model (for LOADEST details see Runkel et al. [8].

LOADEST analysis showed that data from all stations generally fit the models well for all pollutant loads (TN, TP and TSS) as described below. The models were developed for the whole study period at each station. AMLE calibration and estimation option was selected in the models as residuals approximated a normal distribution, and sometimes data were censored. The best models of TN, TP and TSS at each station were selected automatically based on the AIC value in LOADEST. Figure 4 (a, b and c) shows plotting of residuals against the explanatory variables (streamflow and time) and against the predicted value (estimated load) at Coldstream station as a typical case for TN; they were reasonably homoscedastic. Similarly, the goodness of fit in estimation was also tested by normal probability plot of the residuals, and found to be normally distributed, as shown in Figure 4(d). Coefficients of determination (R^2) for the models were evaluated for model performance, and they were greater than 0.85 at all stations for all pollutants, except TP and TSS only at two stations (stations 1 and 10). The R^2 value indicates the "variability explained" by the models for logarithm of load. Figure 5 shows a typical case of comparison between observed and estimated TN loadings (back-transformed to data scale) only on grab sampling days from the sub-catchment at station 4. Observed loads were calculated on the particular grab sampling days by multiplying observed daily streamflow by observed pollutant concentration, and shown in Figure 5 together with estimated loads on those days.

The LOADEST model generates daily loads for each day of the study period, and mean load for the entire study period. Users also may request mean load estimates for seasonal and/or monthly time periods. The daily load values generated by LOADEST were categorized into baseflow loads those fall on the baseflow days and runoff loads those fall on the runoff days in the study period at each station. The numbers of baseflow days and runoff days were calculated previously as per the baseflow contribution criteria and are shown in Table 2. The baseflow loads on all the baseflow days were added up, and then the sum was converted to daily means and then annual values to get mean daily baseflow load and mean annual baseflow load. Similarly, the runoff loads on all the runoff days were added up, and then the sum was converted to daily means and then annual values to get mean daily runoff load and mean annual runoff load. This method has been used in the past by Tortorelli [18], and the same methodology was used in this chapter for consistency. Load separation into these two baseflow conditions and runoff events helps to identify relative load contribution of these two flow regimes to total load at the sub-catchment outlets.

TN, TP and TSS total yields, runoff yields and baseflow yields for the study period at each station were calculated by dividing their mean annual loads by corresponding station drainage area.

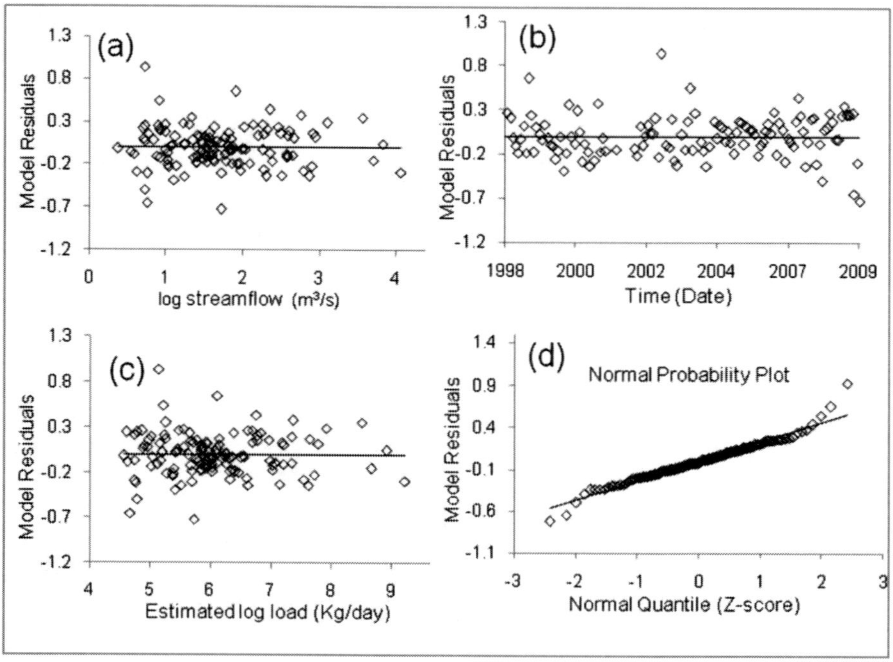

Figure 4. Model residuals against (a) explanatory (b) explanatory and (c) predicted variables, and (d) normal probability plot of the residuals for TN at Coldstream station (station 4).

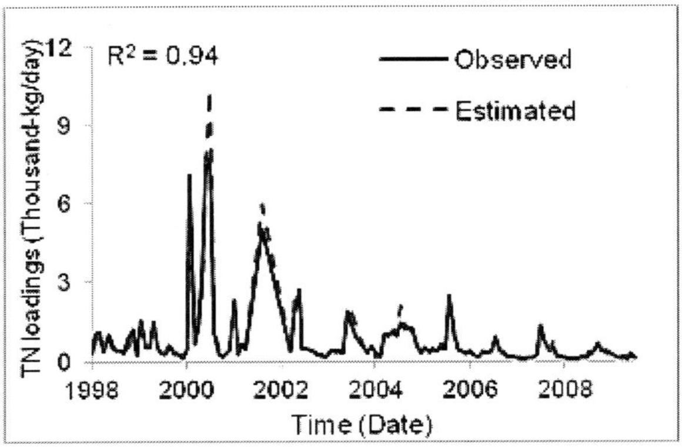

Figure 5. Comparison between observed and estimated TN loadings at Coldstream station (station 4).

4. Results and Discussion

4.1. Pollutant Concentrations

Water quality concentration data obtained through grab samples at each station of the Yarra River catchment were designated as those representing baseflow conditions or runoff events using hydrograph separation. The reason for analyzing and comparing data from these two flow regimes is that different external factors may influence water quality in different flow regimes as discussed in Section 1. The summary statistics of pollutant concentrations categorized into baseflow conditions and runoff events are presented in Table 3. This table shows that a greater number of observed concentrations data (monthly discrete samples) belongs to the baseflow conditions. TN and TP concentrations in both baseflow conditions and runoff events during the study period generally increased from upstream to downstream at main stream stations of the catchment (Table 3). This may be due to the dominant land use type gradually changing from forest to agriculture and then to urban type, from upstream to downstream (Table 2). Moreover, TN and TP concentrations in both baseflow conditions and runoff events were generally greater in tributary stations than in main stream stations. This is because of the drainage areas of tributary stations are explicitly either agricultural or urban types compared to the main stream stations (Table 2). However, TSS concentrations were not significantly different in these cases. At main stream stations, the TN, TP and TSS baseflow mean concentrations were 0.84, 0.04 and 11.60 mg/L respectively, and runoff mean concentrations were 1.23, 0.08 and 32.60 mg/L respectively. At tributary stations, the TN, TP and TSS baseflow mean concentrations were 1.42, 0.09 and 11.86 mg/L respectively, and runoff mean concentrations were 1.79, 0.14 and 36.14 mg/L respectively.

The Mann-Whitney rank-sum test results showed that TN, TP and TSS concentrations were significantly greater ($p \leq 0.01$) in runoff than in baseflow at all stations except TN at stations 2, 6 and 12. Also, the Kruskal-Wallis test results showed that TN, TP and TSS concentrations were significantly greater ($p \leq 0.01$) in urban areas (station11) than in agricultural areas (station 10), and again significantly greater ($p \leq 0.01$) in agricultural areas than in forest areas (station 1).

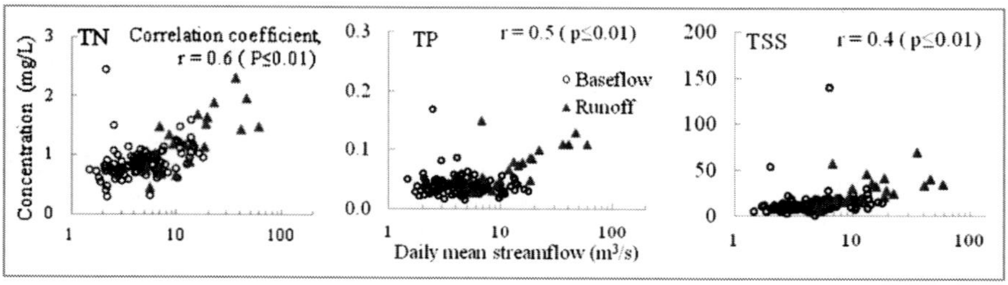

Figure 6. Pollutant concentrations in baseflow conditions and runoff events at Coldstream station (station 4).

Figure 6 shows the pollutant concentrations in baseflow and runoff at station 4 as an example. TN, TP, and TSS concentrations in runoff events generally increased with streamflow at all stations. However, TP concentrations at the main stream and TSS concentrations at any station did not increase with streamflow in case of the baseflow conditions.

Possible causes of larger concentrations of TN, TP and TSS in runoff events than in baseflow conditions are the addition of the pollutants from non-point sources, resuspension of nutrients from the streambed sediment, and stream bank erosion. Wagner and Woodruff [46] also reported similar findings where majority of phosphorus transported in the lake Eucha basin, Oklahoma (USA) was due to non-point sources during runoff events.

Table 3. Summary statistics of TN, TP and TSS concentrations in baseflow conditions and runoff events

Station Number		Pollutant loads	Baseflow concentrations (mg/L)					Runoff concentrations (mg/L)				
			Min.	Median	Mean	Max.	Obs*	Min.	Median	Mean	Max.	Obs*
Main stream	1	TN	0.165	0.511	0.510	0.912	120	0.249	0.723	0.870	3.158	19
		TP	0.003	0.016	0.018	0.110		0.007	0.025	0.044	0.230	
		TSS	1	4	4	10		2	10	32	240	
	3	TN	0.190	0.693	0.705	1.296	113	0.564	1.025	1.143	2.286	26
		TP	0.003	0.027	0.027	0.052		0.017	0.053	0.061	0.300	
		TSS	1	8	9	22		9	23	31	230	
	4	TN	0.306	0.854	0.892	2.454	104	0.457	1.206	1.260	2.306	22
		TP	0.015	0.039	0.040	0.170		0.027	0.070	0.072	0.150	
		TSS	1	12	14	140		6	26	28	69	
	5	TN	0.402	0.976	1.028	2.405	107	0.665	1.280	1.370	2.926	32
		TP	0.009	0.045	0.046	0.079		0.021	0.076	0.089	0.350	
		TSS	1	8	9	24		4	19	37	310	
	9	TN	0.360	1.054	1.062	1.961	112	0.800	1.381	1.482	2.684	37
		TP	0.024	0.066	0.069	0.170		0.042	0.097	0.110	0.360	
		TSS	4	20	22	84		4	33	35	85	
Tributary	2	TN	0.578	1.125	1.140	2.020	115	0.569	1.315	1.373	2.707	24
		TP	0.011	0.033	0.038	0.350		0.017	0.057	0.066	0.140	
		TSS	2	9	10	85		3	16	22	75	
	6	TN	0.376	1.082	1.185	2.439	92	0.020	1.148	1.631	7.861	47
		TP	0.027	0.088	0.090	0.170		0.026	0.110	0.130	0.550	
		TSS	4	15	20	97		5	22	66	670	
	7	TN	0.288	1.059	1.212	3.011	98	0.033	1.276	1.674	5.924	41
		TP	0.028	0.085	0.092	0.190		0.058	0.110	0.127	0.460	
		TSS	4	16	18	66		11	24	55	430	
	8	TN	0.169	0.966	1.106	3.328	87	0.537	1.547	1.615	4.319	35
		TP	0.028	0.083	0.086	0.220		0.049	0.140	0.138	0.240	
		TSS	1	4	6	41		2	12	22	95	
	10	TN	0.202	1.141	1.264	3.460	96	0.428	1.479	1.648	3.530	42
		TP	0.028	0.099	0.103	0.210		0.075	0.135	0.151	0.350	
		TSS	1	4	6	54		1	9	20	100	
	11	TN	0.177	1.543	1.772	13.255	94	0.634	1.931	2.331	5.816	67
		TP	0.049	0.085	0.131	2.600		0.046	0.140	0.201	0.750	
		TSS	1	6	11	180		2	26	42	570	
	12	TN	0.546	2.180	2.274	5.390	101	0.790	2.199	2.236	3.870	38
		TP	0.036	0.096	0.110	0.310		0.035	0.140	0.157	0.490	
		TSS	2	8	12	68		3	16	26	170	

Obs* = number of available observations (or grab samples).

4.2. Estimated Mean Annual Loads and Yields

The summary statistics of mean annual total loads and total yields of pollutants divided into baseflow conditions and runoff events are given in Table 4. In general, mean annual TN, TP and TSS total loads increased from upstream to downstream along the main stream stations of the catchment, with the annual total loads at the outlet station 9 are about 6, 10 and 12 times those of station 1 for TN, TP and TSS respectively. This is because of the drainage area increasing from upstream to downstream, and pollutant loads are proportional to the catchment area. However, station 5 did not follow this trend because of the water pumping out at the Yering Gorge Pumping Station above this station as discussed in Section 3.2 with Table 1. Figure 7 displays the monthly loads of TN, TP and TSS with respect to monthly streamflow at station 4, which shows that pollutant total loads increased with streamflow. Estimated mean annual TN, TP and TSS baseflow loads and runoff loads also increased from upstream to downstream along the main stream stations of the catchment. Figure 8(a) shows that the amount of runoff load contribution into the total load is much higher in the tributary stations than in the main stream stations. Similar observations can be seen in Figure 8(b) for runoff yields. This is expected because tributary stations are either dominantly agriculture or urban types whereas main stream stations are forest-agriculture types (Table 2), and runoff load generation is more influenced by agricultural and urban land use activities. Also, percentages of runoff days were higher in tributary stations than in main stream stations (Table 2). TN, TP and TSS loads delivered during runoff events at all the main stream stations were on average 49%, 54% and 64% of the total loads respectively (Table 4) whereas runoff days were only on average 18% during the study period (Table 2). On the other hand, the runoff loads for TN, TP and TSS at the tributary stations were on average 79%, 81% and 91% respectively (Table 4) whereas runoff days were on average 31% during the study period (Table 2). Also, TN, TP and TSS runoff yields from agriculture dominant areas (station 10) were about 3, 8 and 8 times higher respectively than those of from forest dominant areas (station 1). TN, TP and TSS runoff yields from urban dominant areas (station 11) were about 13, 33 and 42 times higher respectively than those of from forest dominant areas (station 1).

Table 4. Estimated mean annual TN, TP and TSS loads and yields using the regression model LOADEST

Station	Number	Pollutant loads	Mean annual						Load delivered during runoff events (%)
			Total load (kg/yr)	Total yield (kg/yr/ km^2)	Baseflow load (kg/yr)	Baseflow yield (kg/yr/ km^2)	Runoff load (kg/yr)	Runoff yield (kg/yr/ km^2)	
Main stream	1	TN	64000	75.8	37000	43.8	27000	32.0	42
		TP	2400	2.8	1300	1.5	1100	1.3	47
		TSS	1229000	1455.5	402000	476.1	827000	979.4	67
	3	TN	202000	129.0	115000	73.4	87000	55.5	43
		TP	8600	5.5	4500	2.9	4100	2.6	48
		TSS	3501000	2235.1	1767000	1128.1	1734000	1107.0	49
	4	TN	317000	147.2	165000	76.6	152000	70.6	48
		TP	17200	8.0	7500	3.5	9700	4.5	57
		TSS	6324000	2936.7	2768000	1285.4	3555000	1650.8	56
	5	TN	302000	128.3	139000	59.0	163000	69.2	54
		TP	18000	7.6	7000	3.0	11000	4.7	62
		TSS	9531000	4048.1	1773000	753.0	7758000	3295.0	81
	9	TN	406000	122.2	174000	52.4	232000	69.8	57
		TP	28000	8.4	11000	3.3	17000	5.1	60
		TSS	11750000	3536.1	3780000	1137.6	7970000	2398.5	68
Tributary	2	TN	65000	199.7	38000	116.7	27000	82.9	41
		TP	2600	8.0	1200	3.7	1400	4.3	52
		TSS	913000	2804.7	413000	1268.7	500000	1536.0	55
	6	TN	33000	106.8	5000	16.2	28000	90.6	84
		TP	2500	8.1	400	1.3	2100	6.8	86
		TSS	2293000	7419.7	114000	368.9	2179000	7050.8	95
	7	TN	26000	74.1	5000	14.3	21000	59.9	81
		TP	1870	5.3	340	1.0	1530	4.4	82
		TSS	1136000	3239.3	97000	276.6	1039000	2962.7	91
	8	TN	23000	219.6	3000	28.6	20000	191.0	86
		TP	2500	23.9	280	2.7	2220	21.2	89
		TSS	1947000	18590.9	46000	439.2	1901000	18151.7	98
	10	TN	40000	102.0	8000	20.4	32000	81.6	79
		TP	5000	12.7	1000	2.5	4000	10.2	85
		TSS	3292000	8391.5	94000	239.6	3198000	8151.9	97
	11	TN	48000	448.6	4000	37.4	44000	411.2	92
		TP	4900	45.8	300	2.8	4600	43.0	95
		TSS	4414000	41251.9	34000	317.8	4380000	40934.2	99
	12	TN	59000	472.5	9000	72.1	50000	400.4	85
		TP	3900	31.2	500	4.0	3400	27.2	87
		TSS	9948000	79673.0	74000	592.7	9874000	79080.4	99

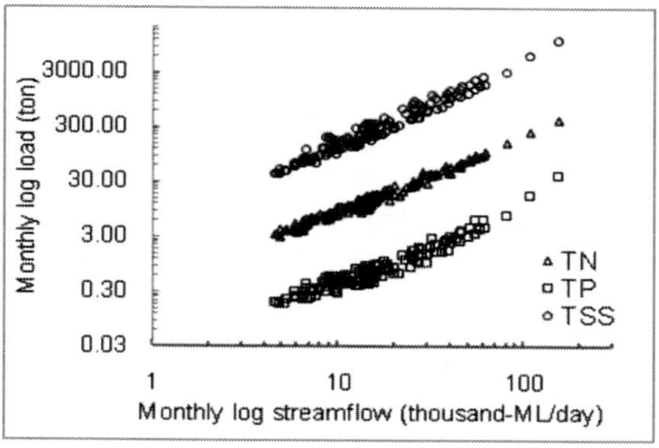

Figure 7. Monthly log loads against monthly log streamflow at Coldstream station (station 4).

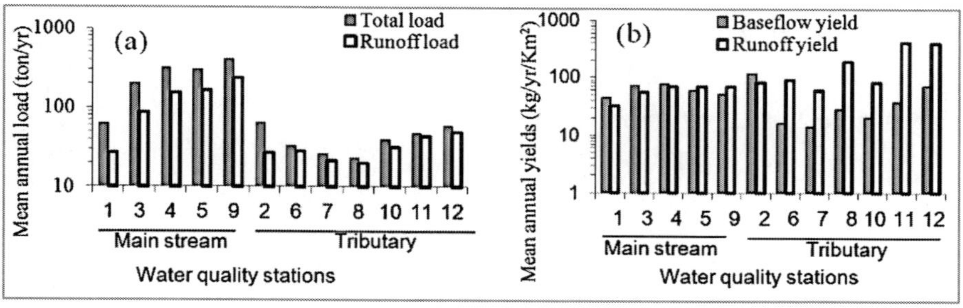

Figure 8. Mean annual (a) total loads and runoff loads, and (b) baseflow yields and runoff yields for TN.

Table 5. Mean annual total loads carried by each segment of the Yarra River catchment

Segment of the Yarra River catchment	Area (Km^2)	Pollutants	Mean annual total load (ton/yr)	Percentage of the total load at catchment outlet
Upper (above station 1; mainly forest dominant area)	844	TN	64	12
		TP	2.4	6
		TSS	1229	4
Middle (between station 1 and 5; mainly agriculture dominant area)	1819	TN	271	49
		TP	18.1	43
		TSS	10595	36
Lower (below station 5; mainly urban dominant area)	1284	TN	218	39
		TP	21.3	51
		TSS	17580	60

Estimated mean annual TN, TP and TSS total loads at the outlet of the catchment (adding up all the loads of stations 9, 10, 11 and 12) were about 553, 42 and 29,404 tons per year respectively. About 65%, 69% and 86% of these annual TN, TP and TSS total loads respectively were delivered by runoff events, although the study period was dry compared to the period of record (Table 1). Moreover, Table 5 shows the loads contribution of the three segments of the study area into the total annual TN, TP and TSS loads at the catchment outlet. It shows that the highest percentage of TN (49%) was generated from the middle agricultural segment, whereas the highest percentages of TP (51%) and TSS (60%) were generated from the lower urban segment. However, the contribution for any pollutant from the upper forest segment to the total annual loads was less than 12%. In all cases, a larger portion of pollutant loads was delivered by runoff events indicating that major pollutant loads were generated from agricultural and urban runoff (i.e., non-point sources).

CONCLUSION

This chapter presents a specific flow regimes based simple procedure to correlate nutrient and sediment loads, and land use activities in a heterogeneous catchment using water quality grab samples, and other readily available data and tools. The proposed procedure addresses the issues of selecting water quality monitoring stations, catchment subdivision, identification of major land uses, analysis of specific flow regimes and estimation of pollutant loads in a heterogeneous catchment. A simple hydrograph separation tool BFP was used to define the specific flow regimes, and a data-based model LOADEST was applied to estimate pollutant loads rather than a complex mechanistic model. Although the procedure does not indicate the specific mechanisms causing surface runoff and pollutant transport, the procedure is simple to apply and leads to good results, particularly in circumstances where limited time (for complex analysis) and limited data are available. The procedure was illustrated in a case study in which nutrient and sediment loads in the Yarra River catchment in Victoria, Australia were assessed.

At all water quality stations analysed, pollutant concentrations were significantly ($p \leq 0.01$) greater in runoff events than in baseflow conditions, and the LOADEST models explained more than 85% variability in pollutant loadings in the Yarra River catchment. The major conclusion from this study was that major pollutant loads were generated from non-point sources during runoff events. Overall, water quality and pollutant concentrations were influenced by flow regimes and land use types, demonstrating the need for water quality monitoring programs to specifically target the collection of water quality samples during runoff events (i.e., storm events) and to characterize the distribution of constituent values in each flow condition. These results are encouraging especially given the pressing need to identify appropriate management practices to improve the water quality within the Yarra River catchment. The presence of increasing nutrient and sediment concentrations, particularly during runoff events, suggests that the Yarra River catchment managers should focus on management of potential non-point sources that occur during precipitation events and surface runoff events in streams.

ACKNOWLEDGMENTS

The authors wish to thank Melbourne Water Corporation, Australian Bureau of Agricultural Resource Economics and Sciences, SILO climate database (Australia), U.S. Bureau of Reclamation, U.S. Geological Survey and NASA-Japan government for providing data and tools for this.

REFERENCES

[1] Ingram, J; Lee, J; Valentin, C. 1996. The GCTE soil erosion network: A multidisciplinary research program. *J. Soil Water Conserv.* Vol. 51(5), 377-380.

[2] ICWE. 1992. The Dublin statement and report of the conference. *International Conference on Water and the Environment: Development Issues for the 21st Century.* Dublin, Ireland.

[3] Heathwaite, A. L; Burt, T. P; Trudgill, S. T. 1990. The effect of agricultural land use on nitrogen, phosphorus and suspended sediment delivery to streams in a small catchment in South West England. In *Vegetation and Erosion.* J. B. Thornes, ed., John Wiley, Chichester, 161–179.

[4] Thomann, R. V; Mueller, J. A. 1987. *Principles of surface water quality modeling and control.* Harper & Row Publishers, New York.

[5] Guo, Y; Markus, M; Demmissie, M. 2002. Uncertainty of nitrate-N load computations for agricultural watersheds. *Water Resour. Res.* Vol. 38 (10), 1-12.

[6] Arnold, J. G; Srinivasan, R; Muttiah, R. S; Williams, J. R. 1998. Large-area hydrologic modeling and assessment: Part I. Model development. *J. Am. Water Resour. Assoc.* Vol. 40 (1), 73-89.

[7] Donigian Jr, A. S; Bicknell, B. R; Imhoff, J. C. 1995. Hydrological simulation program-fortran (HSPF). In *Computer models of watershed hydrology.* V. P. Singh, eds., Water Resources Publications, LLC, Highlands Ranch, Colo, 395-442.

[8] Runkel, R. L; Crawford, C. G; Cohn, T. A. 2004. Load estimator (LOADEST): A Fortran program for estimating constituent loads in streams and rivers. In *Techniques and methods.* US Geological Survey, Reston, Virginia.

[9] Beaulac, M. N; Reckhow, K.H. 1982. An examination of landuse-nutrient export relationships. *J. Am. Water Resour. Assoc.* Vol. 18 (6), 1013-1024.

[10] Vertessy, R. A; Watson, F. G. R; Rahman, J. M; Cuddy, S. M; Seaton, S. P; Chiew, F. H. S; Scanlon, P. J; Marston, F. M; Lymburner, L; Jeannelle, S. 2001. New software to aid water quality management in the catchments and waterways of the south-east Queensland region. *Proceedings of the 3rd Australian Stream Management Conference.* Brisbane, Australia. August 27–29, 611-616.

[11] Worrall, F; Burt, T. P. 1999. The impact of land-use change on water quality at the catchment scale: The use of export coefficient and structural models. *J. Hydrol.* Vol. 221(1-2), 75-90.

[12] Wheater, H. S; Jakeman, A. J; Beven, K. J. 1993. Progress and directions in rainfall-runoff modeling. In *Modeling change in environmental system.* A. J. Jakeman; Beck, M.B; McAleer, M.J., eds., Wiley, New York, 101-132.

[13] Reckhow, K. H; Beaulac, M. N; Simpson, J. T. 1980. *Modeling phosphorus loading and lake response under uncertainty: A manual and compilation of export coefficients,* EPA 440/5-80-011. U.S. Environmental Protection Agency. Washington, D.C.

[14] Hodge, T. A; Armstrong, L. J. 1993. Use of a Multiple Linear Regression Model to Estimate Stormwater Pollutant Loading. In *New techniques for modeling the management of stormwater quality impacts.* W. James, ed., Lewis Publishers, Boca Raton, Florida, 201-214.

[15] Shrestha, S; Kazama, F; Newham, L. T. H. 2008. A Framework for Estimating Pollutant Export Coefficients from Long-term In-stream Water Quality Monitoring Data. *Environmental Modelling & Software.* Vol. 23(2), 182-194.

[16] Newham, L. T. H; Drewry, J. J. 2006. Modelling Catchment Scale Nutrient Generation: Technical Report 28/05, National River Contaminants Program of Land and Water Australia. *CSIRO Land and Water.* Canberra, Australia.

[17] White, K. L; Haggard, B. E; Chaubey, I; USDA, A.R.S. 2004. Water quality at the Buffalo National River, Arkansas, 1991-2001. *Trans. ASAE.* Vol. 47(2), 407-417.

[18] Tortorelli, R. L. 2006. Nutrient concentrations, loads, and yields in the Eucha-Spavinaw Basin, Arkansas and Oklahoma, 2002–2004. *U.S. Geological Survey*. Reston, Virginia.

[19] Wahl, K. L; Wahl T. L. 1995. Determining the flow of comal springs at New Braunfels, Texas. *Texas Water '95*. San Antonio, Texas. 77-86.

[20] USDA-ARS. 1999. *U.S. Department of agriculture-agricultural research service, soil and water assessment tool,* SWAT: Baseflow filter program. (http://swatmodel.tamu.edu/software/baseflow-filter-program).

[21] Murphy, R; Graszkiewicz, Z; Hill, P; Neal, B; Nathan, R; Ladson, T. 2009. Australian rainfall and runoff revision project 7: Baseflow for catchment simulation. *Water Engineering, Institution of Engineers*. Australia. Report P7/S1/004.

[22] White, K. E; Sloto, R. A. 1990. Base flow frequency characteristics of selected Pennsylvania streams. *US Geological Survey*.

[23] Arnold, J. G; Allen, P. M; Muttiah, R; Bernhardt, G. 1995. Automated base flow separation and recession analysis techniques. *Ground Water*. Vol. 33(6), 1010-1018.

[24] Institute of Hydrology. 1980. *Low flow studies*. Institute of Hydrology. Wallingford, U.K. Report No. 1.

[25] Nathan, R. J; McMahon, T. A. 1990. Evaluation of automated techniques for baseflow and recession analyses. *Water Resour. Res.* Vol. 26 (7), 1465-1473.

[26] Gonzales, A. L; Nonner, J; Heijkers, J; Uhlenbrook, S. 2009. Comparison of different base flow separation methods in a lowland catchment. *Hydrol. Earth Syst. Sci.* Vol. 13(11), 2055-2068.

[27] Helsel, D. R; Hirsch, R. M. 1992. *Statistical methods in water resources*. Elsevier, Amsterdam, Netherlands.

[28] IBM. 2010. *IBM SPSS Statistics for Windows*. Version 19.0. NY: IBM Corporation (http://www-01.ibm.com/software/analytics/spss/).

[29] Letcher, R. A; Jakeman, A. J; Merritt, W. S; McKee, L. J; Eyre, B. D; Baginska, B. 1999. *Review of techniques to estimate catchment exports*. Environment Protection Authority. Sydney, NSW.

[30] Marsh, N; Waters, D. 2009. Comparison of load estimation methods and their associated error. *The 18th World IMACS Congress and MODSIM09 International Congress on Modeling and Simulation*. Modeling and Simulation Society of Australia and New Zealand and International Association for Mathmatics and Computers in Simulation, Cairns, Australia, 3332-3328.

[31] Mukhopadhyay, B; Smith, E. H. 2000. Comparison of statistical methods for estimation of nutrient load to surface reservoirs for sparse data set: Application with a modified model for phosphorus availability. *Water Res.* Vol. 34(12), 3258-3268.

[32] Quilbe, R; Rousseau, A. N; Duchemin, M; Poulin, A; Gangbazo, G; Villeneuve, J. P. 2006. Selecting a calculation method to estimate sediment and nutrient loads in streams: Application to the Beaurivage River (Quebec, Canada). *J. Hydrol.* Vol. 326(1-4), 295-310.

[33] Smith, C; Croke, B. 2005. Sources of uncertainty in estimating suspended sediment load. *Symposium S1 Held During the Seventh IAHS Scientific Assembly.* IAHS Press, Foz Do Igaussu, Brazil, 136-143.

[34] Cohn, T. A; Delong, L. L; Gilroy, E. J; Hirsch, R. M; Wells, D. K. 1989. Estimating constituent loads. *Water Resour. Res.* Vol. 25(5), 937-942.

[35] EPA Victoria. 1999. *Protecting the environmental health of yarra catchment waterways policy impact assessment.* EPA Victoria. Australia. Report No. 654.

[36] DSE. 2006. *Strengthening the management of the Yarra and Maribyrnong Rivers: A background report for future water quality management.* Department of Sustainability and Environment, Melbourne.

[37] Davis, J; Breen, P; Hart, B. 1998. *The ecology of the Yarra River: A discussion paper.* Cooperative Research Centre for Freshwater Ecology. Canberra, Australia.

[38] Ng, A. W. M; Perera, B. J. C; Tran, D. H. 2006. Improvement of river water quality through a seasonal effluent discharge program (SEDP). *Water Air Soil Pollut.* Vol. 176(1), 113-137.

[39] Ng, A. W. M; Perera, B. J. C. 2000. Data analysis and preliminary estimates of parameters for river water quality model development. *26th National and 3rd International Hydrology and Water Resources Symposium.* Institution of Engineers, Perth, Australia. 773–778.

[40] Melbourne Water and EPA Victoria. 2009. *Better bays and waterways - a water quality improvement plan for the Port Phillip and Westernport Region: Appendices.* Melbourne Water and EPA Victoria. Australia.

[41] Harris, G; Batley, G; Fox, D; Hall, D; Jernakoff, P; Molloy, R; Murray, A; Newell, B; Parslow, J; Skyring, G; Walker, S. 1996. Port Phillip Bay Environmental Study: Final Report. *CSIRO.* Canberra, Australia.

[42] Winchell, M; Srinivasan, R; Di Luzio, M; Arnold, J.G. 2009. *Arcswat 2.3.4 Interface for SWAT2005, User's Guide.* Grassland, Soil and Water Research Laboratory, Temple, TX.

[43] Jha, M. K; Gassman, P. W; Arnold, J. G. 2007. Water quality modeling for the Raccoon River watershed using SWAT. *Transactions of the ASABE.* Vol. 50(2), 479-493.

[44] Stenback, G. A; Crumpton, W. G; Schilling, K. E; Helmers, M. J. 2011. Rating curve estimation of nutrient loads in Iowa Rivers. *Journal of Hydrology.* Vol. 396(1-2), 158-169.

[45] Akaike, H. 1974. A new look at the statistical model identification. *IEEE Trans. Auto. Cont.* Vol. 19(6), 716-723.

[46] Wagner, K; Woodruff, S. 1997. *Phase I clean lakes project, diagnostic and feasibility study of Lake Eucha.* Oklahoma Conservation Commission, Water Quality Division.

BIOGRAPHICAL SKETCHES

Sushil Kumar Das

Affiliation: College of Engineering & Science, Victoria University, Melbourne, Australia

Education: BSc Eng (Civil), PhD (Civil)

Research and Professional Experience: Dr Sushil Das is a Water/Environmental Engineer with Civil Engineering background. He has excellent academic background, diverse research and around 10 years of working experience in water, environment and other sectors of Civil Engineering, and in university teaching and research. Dr Das's main research interests are water resources planning and catchment management including climate change impact on hydrology and water quality. He has published over 19 referred articles.

Chris Perera

Affiliation: College of Engineering & Science, Victoria University, Melbourne, Australia

Education: BSc (Eng), MSc (Eng), PhD Monash, FIE Aust, CPEng

Research and Professional Experience: Professor Chris Perera is an Emeritus Professor of the College of Engineering & Science at Victoria University. Prior to the Emeritus Professor appointment, he has held several leadership positions in the College and the University, as the Inaugural Dean of the College (2013–2016), as the Associate Dean (Research and Research Training) of the Faculty of Health, Engineering and Science (2008–2012), and the Head of the School of Architectural, Civil and Mechanical Engineering (2002–2008).

Before joining VU in 1992, Professor Perera spent five years as a Postdoctoral Fellow at Monash investigating the flow interaction between rivers and flood plains and optimum operation of multi-reservoir water systems. He also spent five years in the water industry as a hydrologist and a water resources planner both in Australia and overseas. He has also been a visiting research scholar at the International Institute of Applied System Analysis in Austria, the CSIRO and the Indian Institute of Science.

Professor Perera has published over 180 refereed research articles with more than 80 journal papers, and has attracted significant research funding, including from the Australian Research Council. He is also the Editor of International Water Association's Journal of Water and Climate Change. He has supervised over 25 research students, mostly at the PhD level and 5 postdoctoral research fellows. His areas of research are water resources planning (including impacts of climate change and water asset management), urban stormwater management, hydrological modelling, and river water quality modelling.

Professor Perera has provided training and technical support in the past for the REALM water supply planning and management computer software tool, which is widely used in Australia. He also provides specialised consultancy services to water industry in water resources planning, including specialised workshops and teaching into master's level subjects.

Postgraduate Research Supervision:

- Current supervision: 2 PhD students
- Completion: 25 PhD and masters student

Areas of Expertise:

- Urban stormwater management
- Water resources planning (including impacts of climate change)
- Hydrological modelling
- River water quality modelling.

Professional Appointments: Emeritus Professor – College of Engineering & Science, Victoria University, Melbourne, Australia

Honors:

- Vice-Chancellor's Peak Award and Citation for Excellence in Research & Research Training 2011 (Research Team Category)
- Vice-Chancellor's Peak Award and Citation for Excellence in Research & Research Training 2007 (Research Supervision Category)

- Medal for Outstanding Service as Head of School (2002-2006) from staff of School of Architectural, Civil and Mechanical Engineering
- Best Lecturer in Civil Engineering (2006).

Anne WM Ng

Affiliation: College of Engineering & Science, Victoria University, Melbourne, Australia

Education: B Civil Eng (Hons), PhD, Grad Dip Learning and Teaching

Research and Professional Experience: Dr Anne Ng is a Senior Lecturer in Water Engineering and has over 15 years of academic experience at various universities across Melbourne. Dr Ng's main research interest is Water Resources management, including management of Green Infrastructure and Urban Water Demand forecasting. She has published over 70 referred journal and conferences papers.

Areas of Expertise:

- Green Infrastructure
- Water resources management
- River water quality modelling.

Professional Appointments: Senior Lecturer – College of Engineering & Science, Victoria University, Melbourne, Australia

Publications from the Last 3 Years:

1. Imran, H. M.; Kala, J.; Ng, A. W. M.; Muthukumaran, S (2017) An evaluation of the performance of a WRF multi-physics ensemble for heatwave events over the city of Melbourne in southeast Australia, *Climate Dynamics*, 1-34
2. Erabee, I. K.; Ahsan, A.; Jose, B.; Aziz, M. M. A.; Ng, A. W. M.; Idrus, S.; Daud, N. N. N. (2017) Adsorptive treatment of landfill leachate using activated carbon modified with three different methods, *KSCE Journal of Civil Engineering*, 1-13
3. V. M. Jayasooriya, Ng, A. W. M., Muthukumaran, S., Perera, B. J. C. (2017) Optimal Sizing of Green Infrastructure Treatment Trains for Stormwater Management, *Water Resources Management* 30: 5407.
4. Sachindra, DA, Ng, A. W. M., Muthukumaran, S, Perera, B. J. C (2016), Impact of climate change on urban heat island effect and extreme temperatures: A case-study, *Quarterly Journal of the Royal Meteorological Society* 142(694):172-186.
5. Barua S, Ng A. W. M, Muthukumaran S, Roberts P, Perera, B. J. C (2015), Modeling water use in schools: A disaggregation approach, *Urban Water Journal* 1-7.
6. Jayasooriya, V. M. and Ng, A. W. M (2014), *Tools for modeling of Stormwater Management and Economics of Green Infrastructure Practices: a Review*, Vol 225 (8) pp 2055-2075.

In: Nutrients, Wastewater and Leachate
Editor: Amimul Ahsan
ISBN: 978-1-53613-949-5

Chapter 2

THE EFFECT OF NITROGEN RELEASE FROM BIOPLASTIC FERTILIZER (BPF) COMPOSITES: A SLOW RELEASE FERTILIZER CONTROL APPLICATION

A. S. Harmaen*[1,*]*, A. Khalina*[2]*,
***H. Mohd Ali*[3] *and I. Nor Azowa* [4]**

[1]Institute of Tropical Forestry and Forest Products, Universiti Putra Malaysia, UPM Serdang, Selangor. Malaysia
[2]Faculty of Engineering, University Putra Malaysia, UPM Serdang, Selangor. Malaysia
[3]Faculty of Biotechnology and Biomolecular Sciences, Universiti Putra Malaysia, UPM Serdang, Selangor. Malaysia
[4]Faculty of Science, Universiti Putra Malaysia, UPM Serdang, Selangor. Malaysia

[*] Corresponding Author Email: harmaen@upm.edu.my.

Abstract

This paper presents a new technique for using bioplastic materials in the production of bioplastic fertilizer (BpF) composites. For the production of BpF, poly (hydroxybutyrate-co-valerate) was used as the biodegradable polymer materials and blended with the granular ammonium sulphate (NPK) fertilizer and oil palm fibre by using extrusion technique. Leaching of N from NPK and BpF was conducted using column soil study for a period of 24 weeks. The highest nutrient content in the leachates was obtained from NPKC1 and NPKC2 fertilizer samples which was about 80% and 75% of N, respectively. Bioplastic fertilizer composites released about 20% to 30% of N from the soil column after 8 weeks. PHBv/EFB/NPKC2 showed the highest leachate of N (38%) over PHBv/NPKC1 (26%), PHBv/NPKC2 (25%) and PHBv/EFB/NPKC1 (28%). The BpF composites were applied to the oil palm seedlings and its growth performance were evaluated. Plant growth performance in terms of height and diameter were the highest in the oil palm seedlings treated with NPKC1 and NPKC2. For the BpF composites, the higher plant height and diameter were obtained from PHBv/EFB/NPKC2. SPAD readings show that NPKC1 and NPKC2 creates the higher chlorophyll content index to the leaf of oil palm seedlings compared to BpF fertilizer composites.

1. Introduction

Polyhydroxyalkanoates (PHAs) are a family of polyesters produced by bacterial fermentation and has potential in replacing the conventional hydrocarbon-based polymers. Polyhydroxybutyrate-co-valerate (PHBv) is one of the polyhydroxyalkanoate-type polymers. Compared to conventional polyesters, PHBv has much lower crystallinity, decreased stiffness and brittleness, as well as increased tensile strength and toughness. It also has higher melt viscosity, which is a desirable property for extrusion and blow molding [1]. The melting temperatures of the PGBv range from 50 °C to 180 °C, depending on their chemical composition. PHBv has similar material properties to polypropylene, which is good resistance to moisture and aroma barrier properties [2].

As the world second largest producer and exporter of palm oil, Malaysia's palm oil industry leaves behind a huge amount of biomass from replanting and milling activities as compared to other types of biomass. Therefore, biomass from oil palm industry has a great potential to be converted to commercial products such as animal food, fertilizer and absorbent. Oil palm biomass such as empty fruit bunch (EFB) fibers are renewable and readily available resources from palm oil industries and are considered as wastes [3]. This wasted EFB fiber could potentially serves as a reinforcing agent for polymer matrixes. It has emerged as a competitor to the other natural/cellulose fibers [4, 5]. However, natural fibers including the EFB fiber possess a few limitations like moisture affinity, low thermal stability, and poor compatibility with polymer matrix. Although the hydrophilic character of natural fibers would lead to composites with weak interface, natural fiber-based composites are environmentally friendly that worth exploring.

Application of biodegradable polymer as coating of fertilizer granules is the most common controlled release mechanism. This polymer coating offers the best control release over a sustained period of time. Coating the fertilizers with a biopolymer helps immobilize the fertilizer pellets and hence make them resistant to runoff and leaching. Various degrees of leaching resistance have been observed depending on the type of controlled release fertilizer used. The process of fertilizers coating using different kinds of materials and various techniques such as rotating drum, fluidizer bed and spouted bed has been studied widely [6].

Nitrogen (N) is consumed by crop plants as an essential element for plant growth and development. However, N in nitrate form is also highly leachable, particularly in soils and can contaminate groundwater [7, 8]. Slow-release fertilizer (SRF) and controlled-release fertilizer release N, and in some cases, other fertilizer elements, at different rates and through different mechanisms [9]. Microbial decomposition is the primary mechanism of N release, with the carbon in the polymer providing the site for microbial activity. Environmental factors that affect soil microbial activity also affect the N availability (e.g., soil temperature, moisture, pH, and oxygen content) of these products [10]. The coating limits or controls the rate of water penetration to the soluble fertilizer core, and controls the release rate of the solubilized fertilizer from within the granule to the soil. SRF could be a solution to reduce fertilizer nutrient losses, particularly nitrogen leaching, and successively reduced the pollution imposed to the environment. Therefore, this study aims to investigate the feasibility of BpF fertilizer composites as a type of SRF. The leaching of nitrogen of BpF fertilizer composites compounded with PHBv, EFB and NPKC were determined in this study. The effects of different types of BpF fertilizer composites on the growth performance of oil palm seedlings were also evaluated.

2. Materials and Methods

2.1. Materials

Polyhydroxybutyrate-co-valerate (PHBv) under the trade name Biopol was supplied by Innovative Pultrusion Sdn. Bhd., Senawang, Negeri Sembilan, Malaysia. The natural fibre used as fillers were oil palm empty fruit bunch fibres (OPEFB), which were supplied by Poly Composite Sdn. Bhd., Teluk Intan, Perak, Malaysia. Fertilizers, namely NPKC1 (uncoated) and NPKC2 (coated) were supplied by Diversatech Fertilizer Sdn. Bhd., Bangi, Selangor, Malaysia.

2.2. Methods

2.2.1. Bioplastic Compounding Process

For the production of bioplastic fertilizer (BpF) composites, the materials were firstly compounded using a twin-screw extruder with a 30 mm screw diameter. Extruder screw speed was maintained at 50 rpm. BpF composites with various formulation were produced. Table 1 shows the formulation of bioplastic fertilizer.

Table 1. Formulation and production of bioplastic fertilizer (BpF) composites compounding ratio

Composites	Code	PHBv Matrix (%)	EFB Fibre (%)	NPK Fertilizer (%)
NPKC1	C1	0	0	100
NPKC2	C2	0	0	100
PHBv/NPKC1	PH1	40	0	60
PHBv/NPKC2	PH2	40	0	60
PHBv/EFB/NPKC1	PH3	30	10	60
PHBv/EFB/NPKC2	PH4	30	10	60

New techniques for producing bioplastic fertilizer (BpF) composites, namely the extrusion method, were used. Temperature was setting from section I to section II about 170 to 150 °C. The sample BpF composites was extruded from mould in section III (Figure 1). The temperature used for the first and second phase was 170 °C and 150 °C, respectively. The BpF composites were then extruded from mould in the third phase. The BpF composites produced are shown in Figure 1.

Figure 1. Bioplastic fertilizer (BpF) composite samples.

2.2.2. Preparation of Soil for Column Method

The leachate analysis of BpF fertilizer composites were studied using soil column method. Top soil was collected from University Agriculture

Park, UPM, located in Puchong, Selangor, Malaysia (02°N 59.035', 101°E 38.913'). The experimental soil was sandy clay loam. The soil was filled into the 30-mm columns made of acrylic glass before the BpF composites samples were placed inside the columns. NPK and BpF fertilizer composites were incorporated to a depth of 100 mm in the soil columns. A total of six columns were used for the leachate collection of six different BpF composites formulation as shown in Table 1. After the columns were filled, it was saturated with distilled water and were then kept in a controlled temperature room at 30 ± 1 °C. The leachates from each columns were collected after 4, 8, 12, 16, 20 and 24 weeks, respectively, for the determination of nitrogen (N) content. Determination of N content was conducted by using Auto Analysis.

2.2.3. Plant Growth Performance Evaluation

For the plant growth performance, 10 replications of 4-month-old oil palm seedlings were used for each type of BpF composites. BpF composites were applied to the oil palm seedlings and its growth performance in terms of plant height, diameter and leaf cholophyll content index were evaluated within 24 weeks period.

Figure 2. Leachate of BpF Fertilizer Composites using soil column method.

2.2.4. Statistical Analysis

The collected data were statistically analyzed using SPSS 23.0. Analysis of variance (ANOVA) were performed to investigate the effects of types of biofertilizer composites on the growth performance of oil palm seedlings.

3. RESULTS AND DISCUSSION

3.1. Leachate of Nitrogen from Bioplastic Fertilizer Composites

Leachates with the highest nutrient content was observed in the NPKC1 and NPKC2 samples between 12 and 24 weeks (Figure 3). The pattern of nutrient content released from the NPK fertilizers was similar. However, a plot of cumulative nutrient content in the leachate indicated a clear difference between NPKC1 and NPKC2 fertilizer compound. Figure 3 shows that the leaching of N for NPKC1 and NPKC2 fertilizer was higher after 12 weeks. About 25% and 30% of N was released for NPKC1 and NPKC2 fertilizer, respectively. After 24 weeks, NPKC2 showed the highest N leaching, followed by NPKC1 and Control, which were about 80%, 75% and 0.1%, respectively. The NPKC2 compound fertilizer gave higher nutrient release compared with the NPKC1 compound fertilizer.

The PHBv/EFB/NPKC2 composites showed higher N leaching rate compared with other BpF fertilizer composites after 8 weeks. PHBv/EFB/NPKC2 exhibited higher leaching of N owing to the more active degradation on surface and diffusion in NPK and fibre from BpF fertilizer composites. Figure 4 shows that the N leaching of the composites made from PHBv were decreased to 5% after week 12. The BpF composite fertilizer showed no more leaching or released on the following weeks until weeks 24 as observed in Figure 4. Overall, the total mineral N recovery in the leachate of BpF fertilizer composites was low.

In general, the release of N from coated fertilizers is affected by soil temperature which affects diffusion because of nutrient mobility in wet soil. Nitrogen release from polymer coated fertilizer is somewhat dependent on soil temperature. Many reports from researchers revealed that soil temperature has the greatest influence on N release from polymer coated fertilizer [11, 12, 13]. Some of NH_4 may be lost in leachate [14] or may be adsorbed to humus colloids and thus resulting in less leachable N in the water.

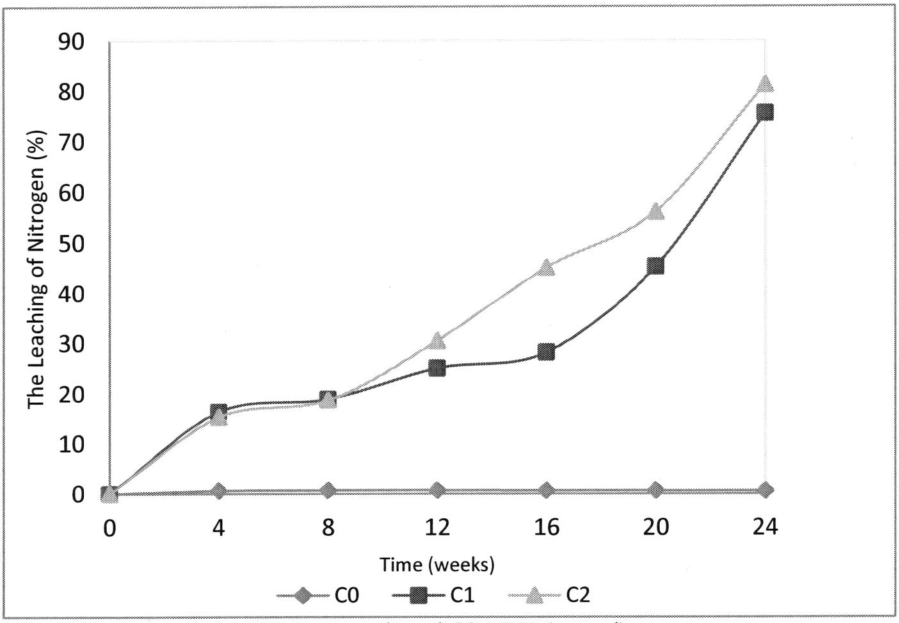

Note: CO-control, C1-NPK (uncoated) and C2-NPK (coated).

Figure 3. The leaching of nitrogen from Control and NPKC2.

3.2. Growth Performance

The BpF composites were applied as slow-release fertilizer to the seedlings of oil palm. This type of fertilizer is expected to only allow small fluctuation in their release within a period of time, ensuring the availability of the nutrients in the soil for continuous plant uptake. However, different formulations of BpF composites may produce varying nutrient availability, therefore growth performance of oil palm seedlings were evaluated to assess the release pattern of SRF.

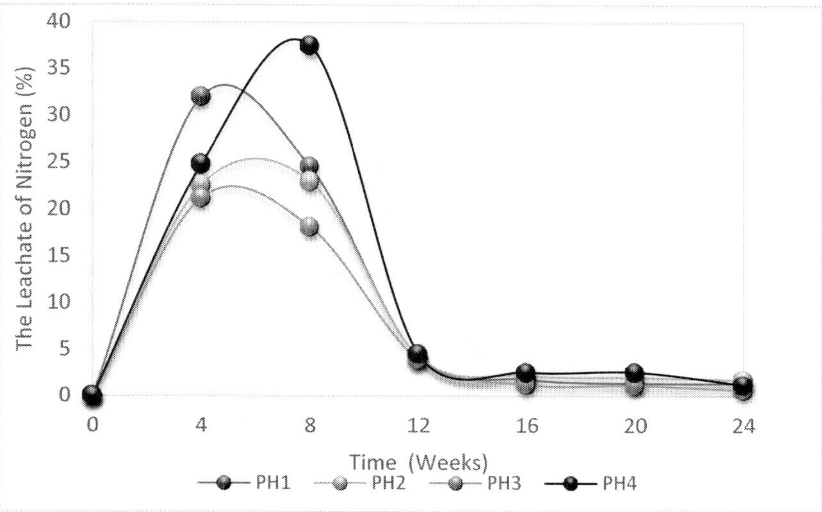

Note: PH1-PHBv/NPKC1; PH2-PHBv/NPKC2; PH3-PHBv/EFB/NPKC1; PH4-PHBv/EFB/NPKC2.

Figure 4. The Leaching of N from BpF Fertilizer Composites.

Table 3. ANOVA analysis of oil palm seedlings growth performance

Source	**DF**	**Anova SS**	**Mean Square**	**F Value**	**Pr >F**
Height of plant	41	35175.98095	857.95075	41.2	<.0001
Dia. of plant	41	15792.05249	385.17201	29.5	<.0001
Chlorophyll of plant	41	7881.23567	192.22526	9.04	<.0001

Table 4. Analysis of oil palm seedlings growth performance after BpF composites application

Treatment	**Height of plant**	**Diameter of plant**	**Number of plant**	**SPAD of plant**
CO	33.80^{e}	16.46^{c}	7.06^{c}	42.01^{d}
C2	45.57^{a}	21.70^{a}	7.69^{a}	53.31^{a}
PH-4	41.91^{bc}	16.46^{c}	7.51^{ab}	49.09^{c}

Note: Means with the same letter are not significantly different at ($p < 0.05$) using Duncan test. CO (Control); C2- NPK coated; PH-4 (PHBv/EFB/NPKC2).

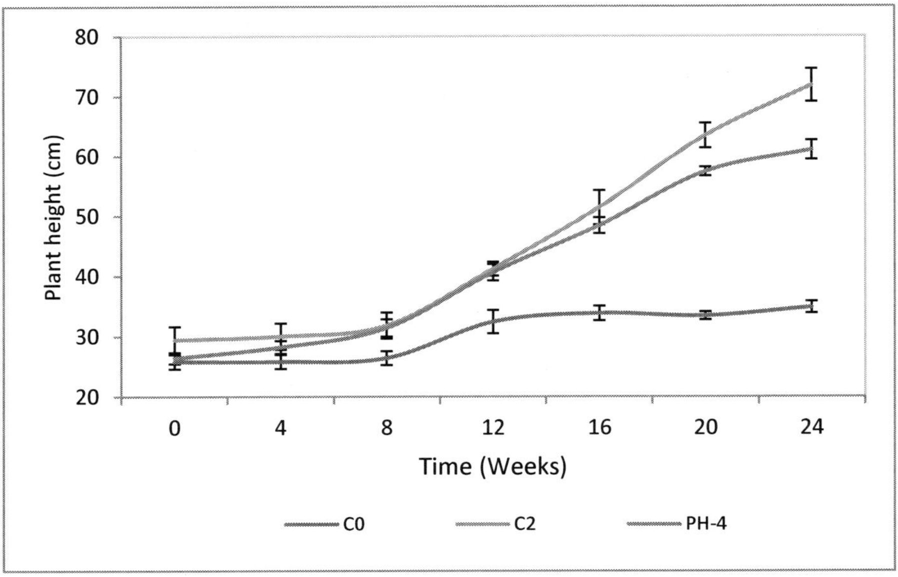

Note: CO-Control; C2-NPK (coated); PH-4 (PHBv/EFB/NPKC2).

Figure 5. Effect of BpF composites on plant height.

Significant differences of oil palm seedlings growth performance were observed after BpF composites application (Table 3 and 4). After 24 weeks of fertilizer application, growth performance of oil palm seedlings increased, indicating that all of the nutrients were released. Based on the analysis of variance (ANOVA) results, the plant's diameter, height and chlorophyll content were significantly influenced by the types of fertilizer applied (Table 3 & Table 4). The differences of the outcome were actually visible through visual observation. Figure 5 shows the effect of fertilizers on plant height. The weight of fertilizer applied in each polybag was about 30 g. After 12 weeks, oil palm seedlings treated with NPK (coated) showed higher rate of plant height compared to those with BpF composites fertilizer. The increasing height was caused by nutrient uptake from the plant root. Oil palm seedlings showed increased growth after nutrient application, which mostly attributed to increased availability of N, P and K. The plant growth was mostly due to the increased leachate of N and the degradation of BpF composites in soil. On the contrary, the control (no fertilizer) showed stunted growth after 12 weeks.

After 24 weeks, the height of oil palm seedlings treated with NPKC2 reached 70 cm, while the height of oil palm seedlings treated with PH-4 reached about 60 cm. The plant height increased because they might uptake more available nutrients in soil.

Based on the empirical results presented in this study, the conceptual model of nutrient release for the formulated fertilizers is explained as follows. The process began with the diffusion of water through the polymer layer following its immersion in water. The water penetrated into the inner granule and solubilised the readily water-soluble fertilizer compounds, dissolving them into the water. After that, the water containing the nutrients were diffused through the soil and uptake by the plant's root. The experiments on nutrient release presented in this study were all conducted in soil. The rate of diffusion controls the rate of nutrient release. In performing the nutrient release experiments of fertilizers coated with biodegradable polymers in soil, other parameters and factors such as microorganism-induced biodegradation should also be considered.

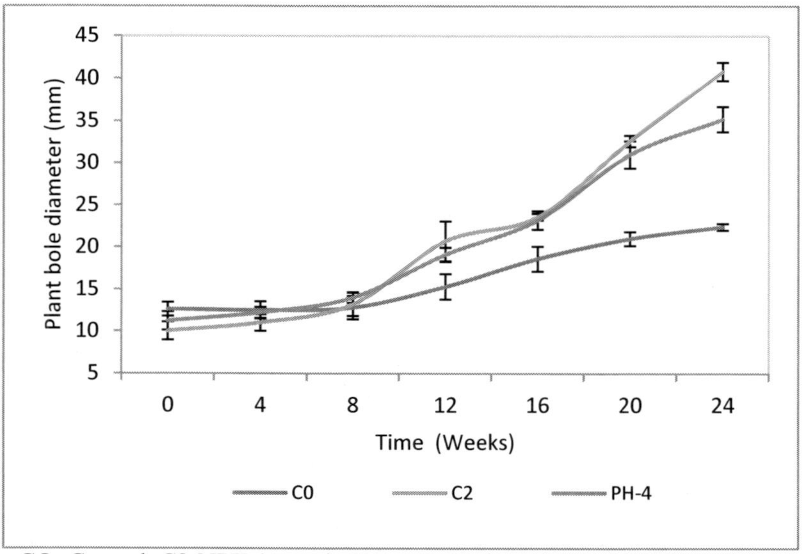

Note: CO- Control; C2-NPK (coated); PH-4 (PHBv/EFB/NPKC2).

Figure 6. Effect of BpF composites on plant bole diameter.

Figure 6 illustrates the results of plant bole diameter measurements in 24 weeks as function of application of different BpF fertilizer composites. Oil palm seedlings treated with NPKC2 exhibited bigger diameters compared to that of the seedlings treated with BpF composites. Due to its slower nitrogen releasing properties, oil palm seedlings treated with BpF composite PH-4 had smaller diameter in comparison to the seedlings treated with NPKC2. However, oil palm seedlings treated with both fertilizers were higher than the control oil palm seedlings.

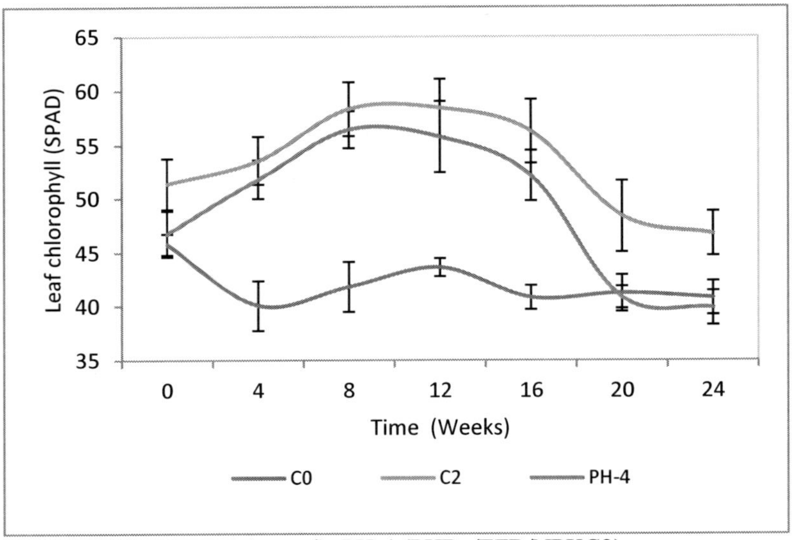

Note: CO-Control; C2-NPK (coated); PH-4 (PHBv/EFB/NPKC2).

Figure 7. Effect of leaf chlorophyll content using BpF composites.

Figure 7 presents the leaf chlorophyll content index in the form of Minolta SPAD-502 values. In comparison to the control oil palm seedlings, leaf chlorophyll content increased after fertilizer application (Figure 7). Brito [15] concluded in his study that portable chlorophyll meter such as SPAD-502 is useful in accurately predicting the concentration of photosynthetic pigments of leaves without involving destructive chemical reagents and complicated laboratory procedures. In this study, oil palm seedlings aged 4 months old were used. Generally, the SPAD threshold for oil palm between the ages of 1–3 years ranges from 55 to 75 [16]. Oil palm seedlings treated with granular NPKC2 displayed higher leaf chlorophyll content index compared to that of the seedlings treated with BpF composites. Slow-release properties of the BpF composites might be the probable explanation for this phenomenon. From Figure 8, one can see that the highest readings of chlorophyll content were recorded in oil palm seedlings treated with NPKC2, which were about 56. The chlorophyll content increased from 4 weeks to 12 weeks but then decreased after 16 weeks. After 16 weeks, the chlorophyll content for oil palm seedlings treated with NPKC2 decreased to about 40. Chlorophyll content index range between 55 and 65 indicated that the plants were moderately healthy. Such outcome was mainly due to that NPK fertilizers were released faster than BpF composites. After 16 weeks, no more nutrients were released and so the plants could not uptake any more nutrients from soil. As a result, the colour of the leaves changed from green to light green. Similar trend was also observed for BpF PH-4 as the chlorophyll content index range increased in the first 12 weeks and decreased afterwards. Control seedlings showed the lowest leaf chlorophyll content index because it did not treat with any fertilizer to provide nutrients for the plant growth.

CONCLUSION

In conclusion, after 24 weeks, NPKC2 showed the highest N leaching, followed by NPKC1 and Control, which were about 80%, 75% and 0.1%, respectively. The NPKC2 compound fertilizer exhibited higher nutrient release compared with the NPKC1 compound fertilizer. PHBv/EFB/NPKC2 composites showed higher leaching of N owing to the more active degradation on the surface and the diffusion in NPK and fibre from BpF fertilizer composites. The leaching of N from PHBv fertilizer composites decreased from 35% to 5% after week 12. PHBv/EFB/NPKC2 composites contain NPKC2 showed higher N leaching rate (about 37%) compared with other BpF fertilizer composites after 8 weeks. Application of granular NPKC1 and NPKC2 recorded higher leaf SPAD index than BpF composites fertilizer compounded with PHBv. The higher chlorophyll content index of about 58 and 56 were recorded in NPKC1 and NPKC2. It can be concluded that, BpF fertilizer composites had slower release pattern compared to that of the commercial fertilize (NPKC1 and NPKC2) and therefore could be potentially used as SRF.

REFERENCES

[1] Hanggi, J. U. 1995. Requirements on bacterial polyesters as future substitute for conventional plastics for consumer goods. *FEMS Microbiology Review*. Vol. 16, 213-220.

[2] McChalicher, S. F. 2007. Investigating the structure–property relationship of bacterial PHA block copolymers. *Journal of Biotechnology*. Vol. 132, 296-302.

[3] Li, Y.; Sun, X. S. 2010. Preparation and characterization of polymer-inorganic nanocomposites by in situ melt polycondensation of L-lactic acid and surface-hydroxylated MgO. *Biomacromolecules*. Vol. 11, 1847-1855.

[4] Rozman, H. D.; Lai, C. Y.; Ismail, H. 2000. The effect of coupling agents on the mechanical and physical properties of oil palm empty fruit bunch polypropylene composites. *Polymer International*. Vol. 49, 1273-1278.

[5] Alam, A. K.; Shubhra, T. H.; Islam, M. R. 2011. Preparation and characterization of natural silk fiber reinforced polypropylene and synthetic e-glass fiber reinforced polypropylene composites: a comparative study. *Journal of Composite Materials*. Vol. 45, 2301-2308.

[6] Morita, A.; Takano, H.; Oota, M.; Yoneyama, T. 2002. Nitrification and denitrification in an acidic soil of tea (Camellia sinensis L.) field estimated by delta N-15 values of leached nitrogen from the soil columns treated with ammonium nitrate in the presence or absence of a nitrification inhibitor and with slow-release fertilizers. *Soil Science Journal of Plant Nutrient*. Vol. 48, 585-593.

[7] Sanchez, C. A. 2000. Response of lettuce to water and nitrogen on sand and the potential for leaching of nitrate-N. *Hort Science*. Vol. 35, 73-77.

[8] Hartz, T. K. 2006. Vegetable production best management practices to minimize nutrient loss. *Hort Technology*. Vol. 16, 398-403.

[9] Sartain, J. B.; Hall, W. L.; Littell, R. C.; Hopwood, E. W. 2004. Development of methodologies for characterization of slow-released fertilizers. *Soil and Crop Science Society Florida Proceedings*. Vol. 63, 72-75.

[10] Shaviv, A.; Mikkelsen, R. L. 1993. Controlled-release fertilizers to increase efficient of nutrient use and minimize environmental degradation: A review. *Fertilizer Research Journal*. Vol. 35, 1-12.

[11] Christianson, C. 1988. Factors affecting N release of urea from reactive layer coated urea. *Fertilizer Research*. Vol. 16, 273-284.

[12] Gandeza, A. T.; Shoji, S.; Yamada, I. 1991. Simulation of crop response to polyolefin-coated urea, I: Field dissolution. *Soil Science Society of America Journal*. Vol. 55, 1462-1467.

[13] Huett, D. O.; Gogel, B. J. 2000. Longevities and nitrogen, phosphorus, and potassium release patterns of polymer-coated controlled-release fertilizers at 30 °C and 40 °C. *Communications in Soil Science and Plant Analysis*. Vol. 31, 959-973.

[14] Singh, R.; Nye, P. E. 1986. A model of ammonia volatilization from applied urea. IV. Effect of method of urea application. *European Journal of Soil Science*. Vol. 37: 9-20.

[15] Brito, G. G.; Sofiatti, V.; Brandão, Z. N; Silva, V. B.; Silva, F. M.; Silva, D.A. 2011. Non-destructive analysis of photosynthetic pigments in cotton plants. *Acta Scientiarum. Agronomy*. Vol. 33, 671-678.

[16] Asyraf, M. A. A. 2011. Review: Construction of greenfeed physiological ruler for oil palms within planting age of 1 year after planting (yap) until 25 years after planting. *Greenfeed Bulletin*. 4.

BIOGRAPHICAL SKETCH

Harmaen Ahmad Saffian

Affiliation: Institute of Tropical Forestry and Forest Products (INTROP),
Universiti Putra Malaysia. Serdang, Selangor.

Education: MSc (Biocomposites Technology), PhD (Biocomposites Technology)

Research and Professional Experinence: Harmaen Ahmad Saffian has completed his PhD from Universiti Putra Malaysia, Serdang, Malaysia. He is the Senior Research Officer of Institute of Forestry and Forest Products (INTROP), UPM. More than 25 years' experience in biocomposites. Many paper of his study related biocomposites material using biopolymer and natural fiber as filler. He PhD studied on biopolymer compounded with fertilizer for slow released control as "Bioplastic Fertilizer" composites. Currently, lignin-based modification compounded with PBS and natural fibre for produces insulating composite materials. He also make research collaboration with others universities such as Kasetsart University, Bangkok, Thailand, Tehran University, Iran and Seoul National University, Republic of Korea.

In: Nutrients, Wastewater and Leachate
Editor: Amimul Ahsan ISBN: 978-1-53613-949-5

Chapter 3

THE PREVALENCE OF ANTIBIOTIC RESISTANT FECAL COLIFORM BACTERIA IN WASTEWATER TREATMENT PLANTS

***Kristína Lépesová*[1]*, Tomáš Mackuľak*[2]**
***and Lucia Bírošová*[1,*]**
[1]Department of Nutrition and Food Quality Assessment,
Faculty of Chemical and Food Technology,
Slovak University of Technology, Bratislava, Slovakia
[2]Department of Environmental Engineering,
Faculty of Chemical and Food Technology,
Slovak University of Technology, Bratislava, Slovakia

ABSTRACT

The development of antibiotic resistance represents a global problem not only in human and veterinary medicine, but also from a socio-

---

[*] Corresponding Author Email: lucyja.birosova@gmail.com.

economic point of view. Excessive use of antibiotics and their dissemination into the environment significantly contributes to the increased prevalence of antibiotic resistant bacterial strains. Wastewater analysis show, that urban wastewater treatment plants (WWTP) receive high concentrations of different pharmaceuticals together with antibiotic resistant bacteria. They can get into the sewerage system from households or farms, but mainly from hospitals and health care facilities. However many antimicrobials are not completely biodegraded in WWTPs, their active metabolites may be extended to the biosphere. The combination of antibiotics in subinhibitory concentrations and the presence of antibiotic resistant bacteria inflowing to the WWTP create favourable conditions for the spread of resistance determinants. Low concentrations of antibiotics generate a selective pressure on wastewater microorganisms resulting in mutations leading to antibiotic resistance. On the other hand resistant bacteria are able to transfer their resistance genes to susceptible species by horizontal gene transfer. Such resistant bacteria can subsequently get into river streams in treated wastewater or into the soil by direct application of stabilized sludge.

Nowadays, the occurrence of coliform bacteria in wastewater and stabilized sludge is widely studied. Coliforms are natural inhabitants of our intestinal microbiota, so they get into the sewerage in feces. However outside of the intestinal tract they may be pathogenic and if they receive resistance genes, they may represent a potential risk for human health.

Thanks to their ability easily to exchange resistance genes, coliforms significantly contribute to the development and spread of phenomenon of antibiotic resistance in the environment. Aim of this study was to compare the prevalence of antibiotic resistant coliform bacteria in influent and effluent wastewater as well as in stabilized sludge from different WWTPs in Slovakia and the Czech Republic. The number of total coliforms in wastewaters inflowing to the WWTPs in Slovakia ranged from 10^4 to 10^5 CFU/mL, while in case of the Czech influents this prevalence was lower (10^3–10^4 CFU/mL). In effluent wastewaters the number of coliforms was reduced more than twice in both countries. The highest values of coliforms were recorded in stabilized sludge with numbers ranging from 10^6 to 10^{12} CFU/g in Slovakia and from 10^7 to 10^8 CFU/g in the Czech Republic. Whereas the majority of coliform bacteria detected in all samples were resistant to β-lactam antibiotic (ampicillin) and gentamicin, influent water and stabilized sludge contained mainly multidrug-resistant strains of coliforms.

Keywords: antibiotic resistance, wastewater, coliform bacteria

Introduction

Emergent contaminants in environmental systems comprise an extensive variety of chemical substances characterized by wide distribution and persistence in the environment due to their massive everyday use for different purposes. Among them, antibiotics (ATB) have received particular attention since the late 90s due to correlations between the development and rapid expansion of ATB resistance as well as their total consumption and occurrence in the environment (Carvalho and Santos, 2016).

ATBs have been widely used to protect human health, reduce diseases or promote growth for livestock and poultry in farms (Chen et al., 2014). After administration to humans and animals, up to 90% of the ATBs can be excreted unchanged via urine and feces into the waste stream (Mutiyar and Mittal, 2014). Because many of ATBs cannot be effectively removed by wastewater treatment processes, municipal, agricultural, and industrial wastewaters represent their major entrance sources into the environment (Chen et al., 2014; Carvalho and Santos, 2016). However, some drugs have long degradation time (the half-time of tetracycline degradation is 300 to 500 days), they can accumulate not only in the environment but also in the food chain. Fedorova et al., (2014) performed monitoring of 32 ATBs in 97 fish samples collected in Europe, Africa and Asia. They study revealed that at least one of all studied ATBs was always present in 15 samples of fishes. The most frequent ATBs found in all positive samples were quinolones (Fedorova et al., 2014). The most serious issue is the transfer of resistant bacterial strains from animals to human. This could be promoted by the presence of ATB residues in foodstuffs of animal origin. For this reason, since 2006, the European Union (EU) has banned the use of ATBs as growth promoters in veterinary medicine and has introduced maximum residue limits of ATBs in food (Dorival-García et al., 2016).

In the past decades, residues of ATBs have been detected in different water matrices, including hospital wastewaters, wastewater treatment plant

(WWTP) effluents, biosolids, soil, surface waters, groundwater, sediments and drinking water (Mutiyar and Mittal, 2014). The detected concentrations generally range from $ng.L^{-1}$ to $\mu g.L^{-1}$ according to the aqueous environment matrices (Carvalho and Santos, 2016). It is well known that ATBs pose a significant risk to environmental and the human health, even at low concentrations (Rodriguez-Mozaz et al., 2015). In some microorganisms, subinhibitory concentrations of antimicrobial agents have increased the horizontal gene transfer as well as a frequency of mutations leading to resistance (Birosova and Mikulasova, 2009; Song et al., 2009). Conditions in WWTPs are also favorable for the proliferation of antibiotic resistant bacteria (ARB) and the antibiotic resistance genes (ARG) (Bouki et al., 2013). The environmnet of the activated sludge is also very suitable for cell-to-cell contact as well as gene exchange, because of high bacterial density and nutrient surplus (Wang et al., 2015).

Numerous studies have reported that urban wastewater and sewage sludge obtain high rates of ARB (Reinthaler et al., 2010). According to these facts, the emission of effluent water as well as sludge represents a key route of release of both ARGs and ARB into the receiving environment. In addition, using a dewatered sludge as land fertilizer also contributes to a high proportion of these contaminants in soil, and the long half-life of ARGs in soil results in a higher persistence of ATB resistance (Wang et al., 2015). On the other hand, the incorrect sewerage sludge disposal from WWTPs increases the transmission risk not only of orofacial infections to humans, but also the zoonotic agents to livestock. Microbial contamination of surface waters may also occur (Dudlova and Juriš, 2014).

Just as Fleming warned, injudicious use of ATBs has led to an antimicrobial resistance crisis in which standard treatments are ineffective and infections persist, prolonging illness and increasing risk of death (Hollis and Ahmed, 2014). Worldwide, the antibacterial resistance has increased dramatically over the past few years and is currently recognized as a major medical challenge in most healthcare settings (Theuretzbacher, 2013). In our study the prevalence of antibiotic resistant coliform bacteria in influent and effluent water as well as stabilized sludge from different WWTPs in Slovakia and the Czech Republic was compared.

Material and Methods

Location and Characterization of WWTPs

Samples of influent and effluent wastewater as well as stabilized sludge were collected from three WWTPs in Slovakia (Bratislava, Petržalka, Piešťany) and in the Czech Republic (Vsetín, Valašské Meziříčí, Zubří).

Two of the Slovak WWTPs are situated in the capital city and the third WWTP is in Piešťany. Bratislava is the most populated city in the Slovak Republic, a cosmopolitan capital with approximately 466,000 inhabitants. Petržalka is considered to be the largest suburb of Bratislava, with the highest population density in Slovakia (125,000 inhabitants) as well as all of Central Europe. Three different WWTPs are receiving wastewater originating from Bratislava, while the wider central part of the city sewerages to the Central Bratislava WWTP. The sewerage leading the effluent to this WWTP is approximately 30 km long. Whereas Petržalka WWTP treats wastewater from the largest housing estate in Bratislava (the length of sewerage is 10 km), wastewater from the peripheral parts of Bratislava is treated in the Devínska Nová Ves WWTP (Mackuľak et al., 2016a). Exact characterization of studied plants was referred in Mackuľak et al., (2014) and Mackuľak et al., (2015). Piešťany is a city located approximately 90 km northeast of Bratislava, which can be characterized as the most important spa city in Slovakia with a current population of 35,000. The technological characteristics of Piešťany WWTP were referred previously in Mackuľak et al., (2017).

WWTPs Vsetín, Valašské Meziříčí and Zubří are situated in the eastern part of the Czech Republic. Vsetín with 27,000 inhabitants is lying on the Bečva River. The city sewer is treated at the Vsetín WWTP, while the sewer system is 1–12 km long. Valašské Meziříčí is related from Vsetín approximately 20 km with a current population of 23,000. The city sewer is inflowing to the Valašské Meziříčí WWTP via 1–15 km long sewer system. The smallest population size of all cities has Zubří (about 5,600), which is located from Vsetín approximately 30 km. The city sewer system and also surrounding municipalities are connected with Zubří WWTP. The sewer system is 1–10 km long (Mackuľak et al., 2016b). Characterization of the Check WWTPs was referred previously in Mackuľak et al., (2017).

Sampling Methodology

The wastewater sampling points were situated on the influent and effluent stream of WWTP. For sample collection during 24 hours in 15 minutes intervals an automatic sampler device was used, while stabilized sludge was taken directly from a chamber filter press. Influent and effluent wastewaters as well as the dewatered sludge samples were immediately transported in sterile falcon tubes to the laboratory for microbiological analysis. Collection of samples took place from 2014 to 2015.

Detection of ATB Resistance in Fecal Coliform Bacteria

After the sample transporting, the detection of total and resistant coliform bacteria content in all of the samples was performed. In the first step, 1 g of stabilized sludge was re-suspended in 10 mL of physiological saline. Subsequently, serial 10-fold dilutions of the stabilized sludge as well as the influent wastewater were prepared and 0.1 mL aliquots were applied on the ATB and ATB-free diagnostic media (Chromocult Coliform Agar-Merck, Germany). In case of the effluent wastewater, the volume of 10 mL through GH polypro membrane (0.2 μm, Pall Corporation, US) was filtered that was placed on selective media supplemented with ATBs. For the resistance detection assessment different ATBs (ampicillin, gentamicin, ciprofloxacin, chloramphenicol and tetracycline) in concentrations defining resistance breakpoints given by European Committee on Antimicrobial Susceptibility Testing (EUCAST) and Clinical Laboratory Standards Institute (CLSI) were used. Colonies of total and resistant coliform bacteria were counted manually after cultivation for 24 hours at 37°C. Each experiment was run in a parallel and repeated three times.

RESULTS AND DISCUSSION

Quantitative Representation of Coliform Bacteria Resistant to ATBs in Wastewater Samples from WWTPs in Slovakia and the Czech Republic

The role of human in creating reservoirs of microbial resistance in the environment cannot be disputed. Since the discovery of ATBs, their use in clinical and veterinary praxis has begun to increase, as well as their subsequent dissemination into the environment. It was estimated that several million metric tons of antimicrobial agents have been released into the biosphere for the first half of the century. ATBs inflowing to the WWTPs in wastewater are not biodegraded and they can consequently spread to the environment (Davies and Davies, 2010).

WWTPs are collecting wastewater from many sources including households, public and health care facilities or other institutions. The number of bacteria in raw wastewater is reduced during primary and secondary treatment process, so their prevalence in the treated effluent is much lower than at the inlet. In this part of our study we have focused on the comparison of the resistant coliform bacteria number in samples of influent and effluent water taken from the Slovak and the Czech WWTPs. *Enterobacteriaceae* belong to the most frequent producers of nosocomial infections in health care facilities as well as infections obtained in the community (Khan et al., 2015). The emergence of antimicrobial resistance among members of this family leads to limited treatment options in cases of infections and has been recognized worldwide over the last decade (Badura et al., 2014). Coliform bacteria form a subgroup of the *Enterobacteriaceae* family characterized as indicator microorganisms for fecal contamination of drinking water and WWTP effluents (Naidoo and Olaniran, 2014).

Table 1 shows the total number of coliform bacteria in 24 hours decanted samples of influent and effluent water from Central Bratislava WWTP, Petržalka WWTP, Piešťany WWTP (Slovakia) as well as from Vsetín WWTP, Valašské Meziříčí WWTP and Zubří WWTP (Czech Republic). The number of coliform bacteria in influent wastewater collected from the Slovak WWTPs ranged from 4.5 to 4.9 log orders of colony forming units (CFU) per milliliter. The highest prevalence of coliforms was detected in the influent wastewater from WWTP Piešťany. The prevalence of coliforms at the influent stream was slightly lower in case of the Czech wastewater samples (from 3.02 to 4.15 log CFU/mL). However, the population connected to the largest WWTP in the Czech Republic is much smaller than in the WWTP Central Bratislava or WWTP Petržalka, the total number of coliforms is only about half of log orders lower.

The use and misuse of ATBs is the most important factor leading to ATB resistance around the world (Garrett, 2015). Since 2001, European recommendations and national plans have aimed to improve the management of ATBs (Lefebvre et al., 2015). In 2015, the European Union/European Economic Area (EU/EEA) population-weighted mean consumption of ATBs for systemic use in the community expressed as defined daily doses (DDD) was 22.4 per 1 000 inhabitants and per day (ECDC, Stockholm 2016). According to the European Food Safety Authority surveillance, the use of ATBs was higher among countries with low levels of educations and in worse economic circumstances (European Commission, 2016). However, the average consumption of ATBs in Slovakia in 2015 was higher as the EU mean value (24.5 DDD per 1 000 inhabitants and per day), since 2013 the ATB consumption rate have decreased by 6 percentage points in 2016 (European Commission, 2016; ECDC, Stockholm 2016).

The spread of multidrug-resistant (MDR) bacteria is a public health problem worldwide (Lefebvre et al., 2015). Although many of ARB are not pathogenic, resistance genes can be transferred to pathogens that represents a public health risk (Palacios et al., 2017).

Table 1. The total number of coliform bacteria in wastewater and stabilized sludge samples from WWTPs

Country	WWTP	Influent water log CFU/mL	Effluent water log CFU/mL	Stabilized sludge log CFU/g
Slovakia	Central Bratislava	4.5	1.85	9.91
	Petržalka	4.5	1.56	8.56
	Piešťany	4.94	1.91	10.52
Czech Republic	Vsetín	4.15	2.2	6.7
	Valašské Meziříčí	4.05	1.8	8.1
	Zubří	3.02	0.6	7.8

Several studies have demonstrated the growth in fatal outcomes from bacteremia caused by MDR bacterial infections, such as methicillin-resistant *Staphylococcus aureus* (MRSA), resistant *Pseudomonas aeruginosa*, extended-spectrum β-lactamases (ESBL)-producing *Enterobacteriaceae* and vancomycin-resistant *Enterococcus* (VRE), compared with drug-sensitive bacteria (Picot-Guéraud et al., 2015). Carbapenems are considered to be the only therapeutic options available to treat the serious infections caused by MDR *Enterobacteriaceae* producing ESBLs, but the significantly increased rates in isolated carbapenem-resistant *Enterobacteriaceae* worldwide are of great concern. The *Enterobacteriaceae* isolates producing ESBLs or carbapenemases are always resistant to multiple ATB classes, since the genes encoding ESBLs and carbapenemases are reported to be located on plasmids which often carry resistance determinants against other antimicrobial agents such as aminoglycosides and fluoroquinolones, too (Zhang et al., 2014).

Figure 1 compare the proportion of resistant coliform bacteria at the inlet point of the monitored Slovak WWTPs. High percentage of ampicillin-resistant coliforms with minimal differences between the EUCAST and CLSI resistance breakpoints in all WWTPs was observed. In

Petržalka WWTP, almost 99% of resistant strains showed resistance to this ATB.

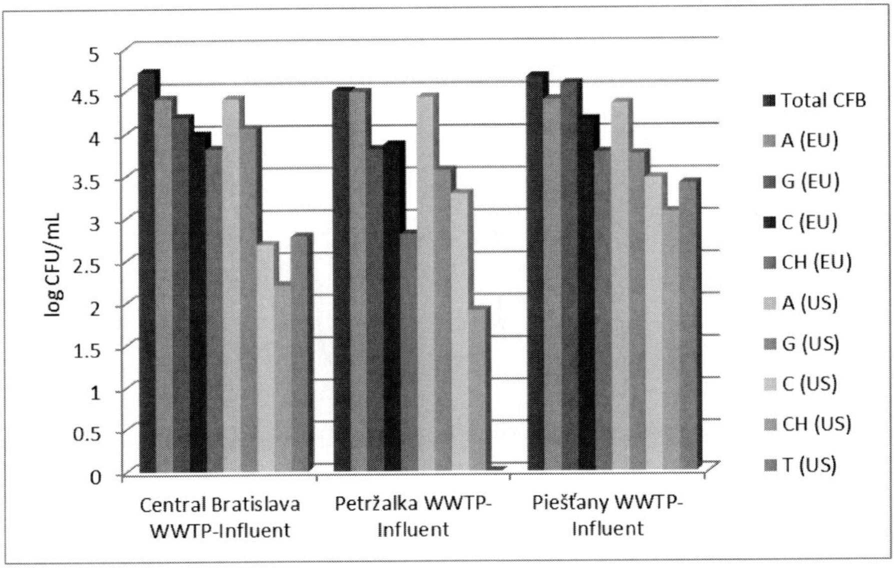

CFB-coliform bacteria, EU-resistance breakpoints defined by EUCAST, US-resistance breakpoints defined by CLSI.

A-ampicillin, G-gentamicin, C-ciprofloxacin, CH-chloramphenicol, T-tetracycline.

Figure 1. The prevalence of resistant coliform bacteria in influent wastewaters collected from WWTPs in Slovakia (log CFU/mL).

It is not surprising, because of intrinsic resistance of coliforms to β-lactam ATBs due to the presence of a chromosomal *ampC* gene. More than 80% of bacterial strains were resistant also to gentamicin and ciprofloxacin according to EUCAST resistance breakpoints. When using ATBs in concentrations determined by CLSI standards a slight decrease in the number of gentamicin- and ciprofloxacin-resistant coliforms was observed. Both of these substances can be found in wastewaters in higher concentrations as a consequence of their frequent use in community and hospital sector (ECDC, Stockholm 2016). The lowest number of coliforms showed resistance to chloramphenicol (<80%) and tetracycline (<70%) according to EUCAST and also CLSI resistance breakpoints. Despite the fact that Piešťany WWTP serves for the smallest population (35 000 p. e.)

among the studied treatment plants, the highest prevalence of coliform bacteria resistant to all tested ATBs in this influent water sample was reported. In addition, more than half of resistant strains survived ATB concentrations given by EUCAST and CLSI. One of the reasons could be that Piešťany is a most popular spa city in Slovakia visited also by patients carrying also resistant bacterial strains. Contrary, the situation was the best in case of Petržalka WWTP influent water, in which no tetracycline-resistant coliforms were detected. However, the percentage of resistant strains was above 50% in all monitored WWTPs suggesting the presence of MDR coliform bacteria.

The average consumption of ATBs in the Czech Republic in 2015 was 19.6 DDD per 1000 inhabitants and per day. This value is lower not only as the EU mean value (22.4 DDD per 1 000 inhabitants and per day) but as the Slovakia mean value (24.5 DDD per 1000 inhabitants and per day), however the ATB consumption rate was showing a slightly increasing tendency between 2011 and 2015. Among all monitored ATB classes in 2015 penicillins, cephalosporins, macrolides, lincosamides and streptogramins were frequently used in the community settings. Compared with Slovakia, in 2015 the consumption of quinolones was reduced more than twice (ECDC, Stockholm 2016). Figure 2 shows resistant coliform bacteria content of influent wastewaters taken from the Czech WWTPs. Similarly to the Slovak influent wastewater samples, a high ampicillin resistance rate (99%) in coliforms from all of the Czech WWTPs was observed, which is also in proportion with the consumption of penicillins in this country. It is important to point out that a group of the total coliform bacteria monitored in wastewaters may include also strains of *Escherichia coli*, which are susceptible to β-lactams (Magiorakos et al., 2012). The number of resistant coliforms was almost comparable in Valašské Meziříčí WWTP and Vsetín WWTP, while wastewater inflowing to the Zubří WWTP contained only ampicillin-, gentamicin- and ciprofloxacin-resistant coliform bacteria. High resistance levels of coliforms to gentamicin in influent wastewaters from Vsetín WWTP (85%), Valašské Meziříčí WWTP (78%) and Zubří WWTP (73%) according to EUCAST resistance breakpoints were observed, while they survived ATB concentrations determined by CLSI. Even though the consumption of quinolones in the Czech Republic is lower than in Slovakia, we have observed high prevalence of ciprofloxacin-resistant coliforms in Vsetín and Valašské Meziříčí WWTPs (80-90%). Their growth was inhibited by concentration of this ATB exceeding EUCAST resistance breakpoints only in Zubří WWTP. While in the sample of influent water from Vsetín WWTP were present coliform bacteria resistant to chloramphenicol according to EUCAST (67%) as well as CLSI (30%). On the contrary, in the influent water from Zubří WWTP no coliforms resistant to chloramphenicol either tetracycline were detected.

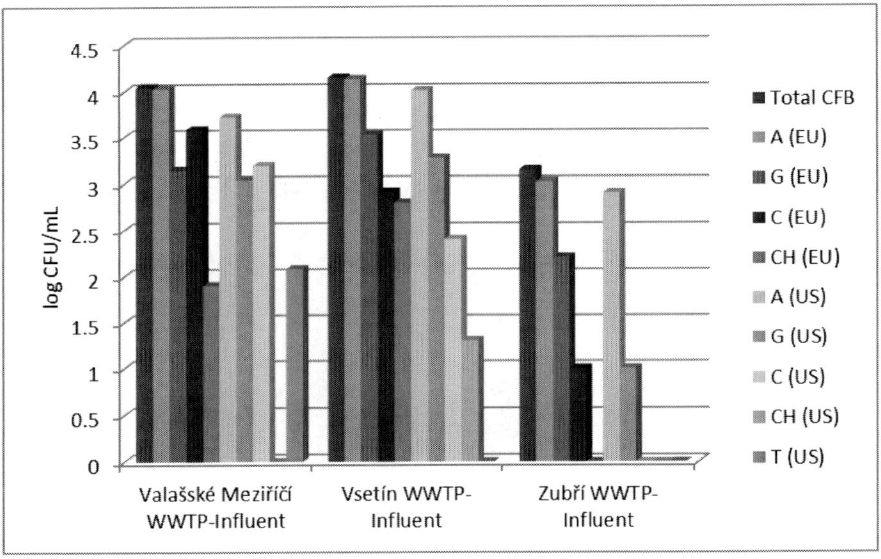

CFB-coliform bacteria, EU-resistance breakpoints defined by EUCAST, US-resistance breakpoints defined by CLSI.

A-ampicillin, G-gentamicin, C-ciprofloxacin, CH-chloramphenicol, T-tetracycline.

Figure 2. The prevalence of resistant coliform bacteria in influent wastewaters collected from WWTPs in the Czech Republic (log CFU/mL).

During the process of wastewater treatment, major qualitative and quantitative changes in the distribution of the bacterial population may occur. Generally, wastewater treatment determines a marked reduction in the bacterial numbers, including the total numbers of resistant bacteria. However, some studies suggest that antimicrobial resistant and susceptible populations are not equally affected (Guardabassi et al., 2002). According to Rizzo et al., (2013) the final effluent of WWTP can discharge about 10^9–10^{12} CFU per day, per inhabitant equivalent bacteria; among these, at least 10^7–10^{10} could have any kind of acquired ATB resistance (Rizzo et al., 2013). For this reason, the monitoring of the prevalence of resistant coliform bacteria in effluent wastewater samples from the same WWTPs in Slovakia and the Czech Republic was performed.

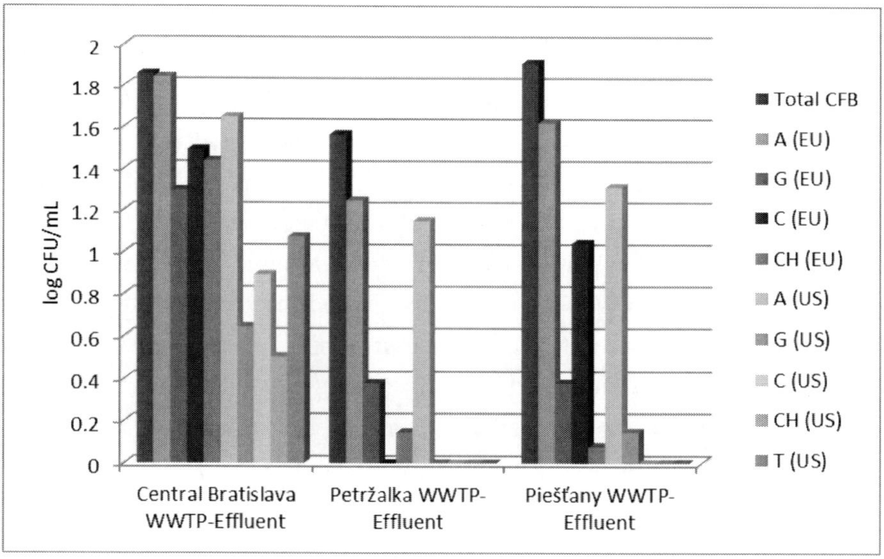

CFB-coliform bacteria, EU-resistance breakpoints defined by EUCAST, US-resistance breakpoints defined by CLSI.

A-ampicillin, G-gentamicin, C-ciprofloxacin, CH-chloramphenicol, T-tetracycline.

Figure 3. The prevalence of resistant coliform bacteria in effluent wastewaters collected from WWTPs in Slovakia (log CFU/mL).

The prevalence of coliform bacteria in Slovak effluent water samples ranged from 1.56 to 1.9 log CFU/mL, which is a significantly lower content than in the influent water samples (Table 1). The highest prevalence of coliform bacteria was determined in the effluent water from Piešťany WWTP, and the lowest in the effluent from Petržalka WWTP. The elimination rate of studied bacteria was so variable among the Czech WWTPs. The highest number of coliform bacteria was recorded in the effluent water from Vsetín WWTP (2.2 log CFU/mL), while their presence in Zubří WWTP was markedly reduced (Table 1).

Figure 3 documents the percentage prevalence of ATB resistant coliform bacteria in effluents from Slovakia. In all samples the large proportion of ampicillin-resistant strains according to EUCAST as well as CLSI was also recorded.

When comparing the WWTPs, the highest number of resistant coliforms was detected in the effluent water of the largest WWTP in Central Bratislava. In addition, up to 50% of presenting strains showed resistance to all tested ATBs according to both EUCAST and CLSI breakpoints indicating MDR character. The Bratislava Water Company states that the purified water is subsequently discharged into the Danube River in a volume of 0.5 m^3 /s, what represents potential risk of contamination and spread of bacterial resistance into the environment (Bratislava Water Company, 2013). Previously, some studies showed that ARB in municipal WWTP effluents discharged into natural water streams may induce the spread of ARB in the environment. Their ability to transmit resistance to other bacteria including pathogens through transmissible resistance factors is an important concern all in the world (Huang et al., 2016). The most significant reduction in the number of resistants was observed in Petržalka WWTP effluent, when we were able to detect only ampicillin-, gentamicin- and chloramphenicol-resistant strains of coliforms according to EUCAST resistance breakpoints. Concentrations of ATBs defined by CLSI have suppressed the growth of coliform bacteria showing resistance to all tested ATBs except ampicillin. Nevertheless the influent wastewater sample taken from Piešťany WWTP contained the highest number of MDR coliform bacteria, after the treatment process except for ampicillin, a low rate of gentamicin, ciprofloxacin and chloramphenicol resistance was observed (EUCAST).

Resistance to all tested ATBs in the effluent water from Valašské Meziříčí WWTP was recorded, while the highest number of coliforms was resistant to ampicillin and gentamicin (Figure 4). Even though Vsetín WWTP effluent water contained the highest number of total coliforms, they were resistant only to ampicillin (EUCAST and CLSI), gentamicin

(EUCAST) and chloramphenicol (EUCAST). On the other hand, the percentage as well as the nominal proportion of resistant coliforms was higher as in case of Valašské Meziříčí WWTP. No resistant strains of coliform bacteria in the effluent sample collected in Zubří WWTP were detected.

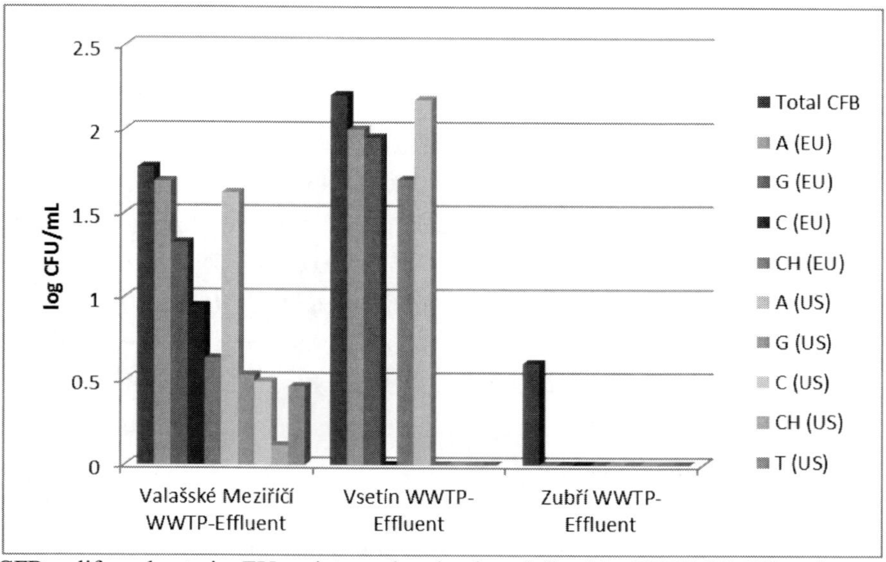

CFB-coliform bacteria, EU-resistance breakpoints defined by EUCAST, US-resistance breakpoints defined by CLSI.

A-ampicillin, G-gentamicin, C-ciprofloxacin, CH-chloramphenicol, T-tetracycline.

Figure 4. The prevalence of resistant coliform bacteria in effluent wastewaters collected from WWTPs in the Czech Republic (log CFU/mL).

The Prevalence of Resistant Coliform Bacteria in Stabilized Sludge from WWTPs in Slovakia and the Czech Republic

Beside wastewater, the most hazardous environmental source of resistance dissemination is a stabilized sludge produced in WWTPs. It's use is primarily preferred in the field of agricultural production because of the high nitrogen, phosphorus, calcium, magnesium, sulfur and organic carbon content. All of these components have considerable importance for composting process (Maulini-Duran et al., 2013; Valderrama et al., 2013). However, in some cases sludge does not have to meet quality requirements for agricultural use. Pollutants that are toxic not only to humans but also show terrestrial as well as aquatic ecotoxicity include heavy metals and organic pollutants like pharmaceuticals, drugs and personal care products (Valderrama et al., 2013; Healy et al., 2016). In addition, sewage sludge may contain pathogenic microorganisms as well as ARB, too (Reinthaler et al., 2010; Maulini-Duran et al., 2013).

In the EU, production of sewage sludge reach approximately 11.5 million tons dry weight/year, up to 70% of which is re-used in agriculture or is disposed of through composting, land filling or incineration (Stiborova et al., 2017). Among the largest sludge producers are ranked Germany (2.2 million t/ year), United Kingdom (1.8 million t/ year), Spain (1.1 million t/ year), France (1.06 million t/ year) and Italy (1.05 million t/ year), while Slovakia as well as the Czech Republic is characterized with lowest production rate (Kelessidis and Stasinakis, 2012). Annual production of stabilized sludge in Slovakia is about 58 000 t, which is much lower compared to the Czech Republic (172 000 t) (Kelessidis and Stasinakis, 2012; Birošová et al., 2014). Regarding disposal methods of sludge matrices, a significant decreasing trend in agricultural reuse in these countries from 2000 to 2009 was observed. In 2007 and 2008 the European Statistical System have also reported that about 67% of produced stabilized sludge in Slovakia was applied in the form of compost, while in case of the Czech Republic this percentage was slightly lower (about 40%) (Kelessidis and Stasinakis, 2012). In 2012, up to 36 832 tons of sludge was processed to compost and only 1139 tons of sludge from WWTPs in Slovakia was applied directly to agriculture soil (Birošová et al., 2014).

Because the presence of resistant bacteria in stabilized sludge represent a real threat, our aim was also the quantitative determination of coliform bacteria showing resistance to ATBs in samples of stabilized sludge collected from WWTPs.

While the number of fecal coliforms in Slovak WWTP sewage sludge ranged from 8.6 to 10.5 log CFU/g, their prevalence in samples of stabilized sludge from the Czech Republic was lower (6.7-8.1 log CFU/g)

(Table 1). The highest number of coliform bacteria was detected in the sludge sample from Piešťany WWTP (10.5 log CFU/g) and resistance to all tested ATBs according to EUCAST and CLSI resistance breakpoints was observed (Figure 5). More than 50% prevalence of resistant coliforms in sludge sample indicates that the majority may be considered as MDR. Although, the stabilized sludge from Petržalka WWTP was characterized by the lowest number of coliform bacteria, the proportion of resistant strains was almost the same as in case of sludge from the Central Bratislava WWTP. Coliforms able to survive in the presence of all tested ATBs according to EUCAST as well as CLSI were detected in both sludge samples.

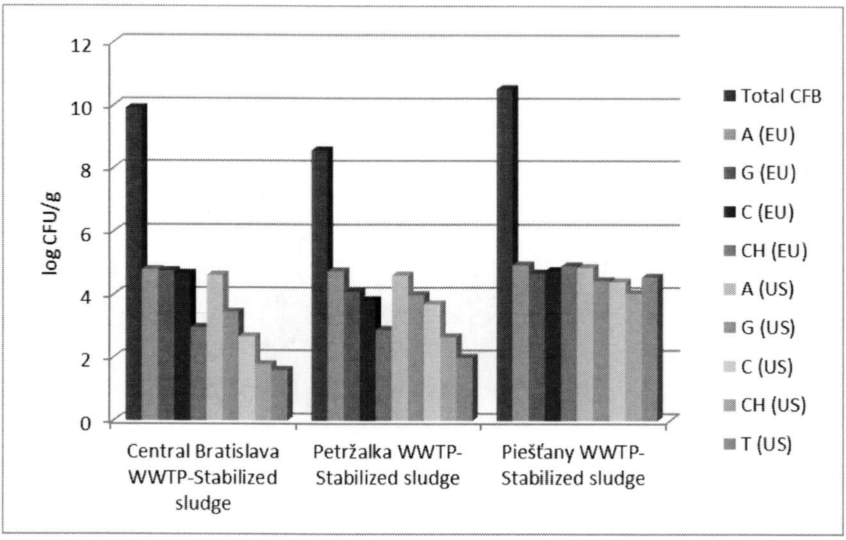

CFB-coliform bacteria, EU-resistance breakpoints defined by EUCAST, US-resistance breakpoints defined by CLSI.

A-ampicillin, G-gentamicin, C-ciprofloxacin, CH-chloramphenicol, T-tetracycline.

Figure 5. The prevalence of resistant coliform bacteria in stabilized sludge collected from WWTPs in Slovakia (log CFU/g).

Among all samples of stabilized sludge from the Czech Republic, the highest prevalence of coliform bacteria in Valašské Meziříčí WWTP was observed. Simultaneously, more than half of present coliforms were resistant to all tested ATBs according to the resistance breakpoints determined by EUCAST and CLSI (Figure 6). This suggests the prevalence of MDR fecal coliform content in this sludge sample. The situation in the stabilized sludge taken from Vsetín WWTP was almost identical; however the total number of coliforms was the lowest. Comparing the prevalence of resistant coliform bacteria in the stabilized sludge and the effluent water from Zubří WWTP, it can be concluded that resistants were attached already into the sludge. In addition, resistant strains detected in this sludge sample were showing MDR character. Higher level of gentamicin-resistant coliforms was also observed.

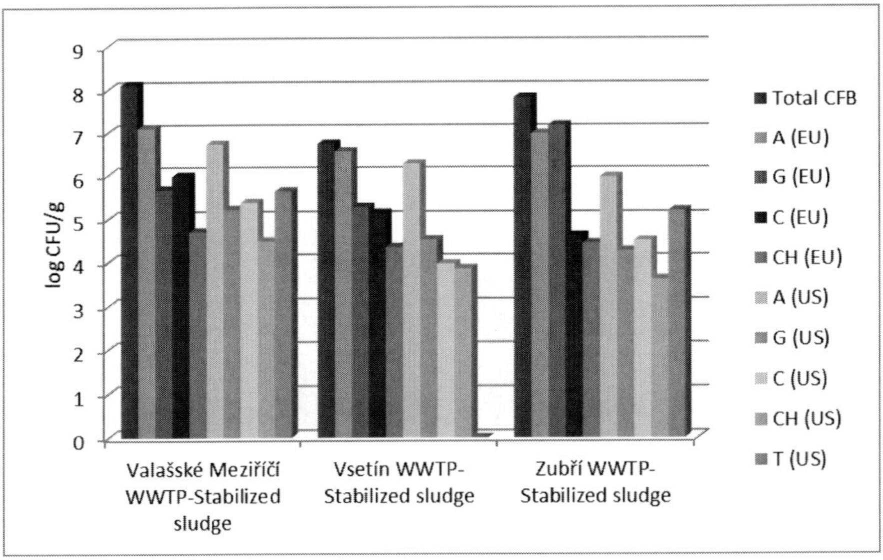

CFB-coliform bacteria, EU-resistance breakpoints defined by EUCAST, US-resistance breakpoints defined by CLSI.

A-ampicillin, G-gentamicin, C-ciprofloxacin, CH-chloramphenicol, T-tetracycline.

Figure 6. The prevalence of resistant coliform bacteria in stabilized sludge collected from WWTPs in the Czech Republic (log CFU/g).

Wastewater treatment processes do not ensure complete elimination of pathogenic microorganisms. These bacteria together with ARB can re-enter the food chain by application of sewage sludge to arable land. Worldwide problem of antibiotic resistance is constantly getting worse in health services as well as in community. This is also evidenced by an increasing prevalence of ESBL-producing pathogens in sewage sludge. Reinthaler et al., (2010) monitored different samples of stabilized sludge collected in Austria. Their results showed that ESBL-producing *E. coli* were ubiquitous in the WWTP sludge (Reinthaler et al., 2010). Birošová et al., (2014) have also monitored the prevalence and seasonal occurrence of resistant coliforms in stabilized sludge. High number of coliforms resistant to ampicillin and gentamicin was detected in samples of stabilized sludge taken from two different WWTPs in winter. In the summer sludge sample, the number of bacteria with high level resistance to all tested ATBs (ampicillin, gentamicin, ciprofloxacin, chloramphenicol and tetracycline) was increased (Birošová et al., 2014).

CONCLUSION

The present study focused on the quantitative prevalence of resistant fecal coliform bacteria in influent and effluent wastewaters as well as in stabilized sludge collected from WWTPs in Slovakia and the Czech Republic. The highest number of coliforms was detected in the influent water from Piešťany WWTP (Slovakia), while the majority of them were resistant to all tested ATBs. Even though wastewater treatment has achieved a reduction in the number of total coliforms up to 3 logarithmic orders, the effluent water from this WWTP still contained ampicillin-, gentamicin-, ciprofloxacin- and chloramphenicol-resistant bacterial strains. While coliform bacteria detected in the effluent water from the Central Bratislava WWTP showed MDR character, the effluent of Petržalka WWTP contained only ampicillin-, gentamicin- and chloramphenicol-resistant coliforms. However, the prevalence of studied bacteria in influent as well as effluent wastewater samples from the Czech Republic was lower compared to Slovakia, the percentage of coliforms showing resistance to ATBs was almost the same. An exception was the effluent water collected from Zubří WWTP which was ATB resistant coliforms free.

All of the studied wastewater and stabilized sludge samples were characterized by a high prevalence of ATB resistant fecal coliforms showing MDR. Based on our results, it can be also concluded that the majority of microorganisms flowing into WWTPs are partially attached to solid matrices such sludge and physically or chemically inhibited during separate treatment technological processes at WWTPs.

Acknowledgments

This work was supported by the Scientific Grant Agency VEGA (Grant number VEGA 1/0096/17), by the Slovak Research and Development Agency (Grant number APVV-16-0171) and by the STU Grant scheme for Support of Young Researchers (contract no. 1686).

References

Badura, A., Luxner, J., Feierl, G., Reinthaler, F. F., Zarfel, G., Galler, H., Pregartner, G., Riedl, R., Grisold, A. J., 2014. Prevalence, antibiotic resistance patterns and molecular characterization of *Escherichia coli* from Austrian sandpits. *Environ. Pollut.* 194, 24-30.

Birošová, L., Mackuľak, T., Bodík, I., Ryba, J., Škubák, J., Grabic, R., 2014. Pilot study of seasonal occurrence and distribution of antibiotics and drug resistant bacteria in wastewater treatment plants in Slovakia. *Sci. Total Environ.* 490, 440-444.

Birosova, L. & Mikulasova, M., 2009. Development of triclosan and antibiotic resistance in *Salmonella enterica* serovar Typhimurium. *J. Med. Microbiol.* 58 (4), 436-441.

Bouki, CH., Venieri, D., Diamadopoulos, E., 2013. Detection and fate of antibiotic resistant bacteria in wastewater treatment plants: A review. *Ecotox. Environ. Safe.* 91, 1-9.

Bratislava Water Company, 2013. *Tlačové správy. Modernizácia dvoch najväčších čistiarní na Slovensku.* [*Press news. Modernization of the two largest wastewater treatment plants in Slovakia.*] http://www.

bvsas.sk/sk/press/tlacove-spravy/modernizacia-dvoch-najvacsich-cistiarni-slovensku.html.

Carvalho, I. T. & Santos, L., 2016. Antibiotics in the aquatic environments: A review of the European scenario. *Environ. Int.* 94, 736-757.

Davies, J. & Davies, D. (2010). Origins and Evolution of Antibiotic Resistance. *Microbiol. Mol. Biol. R.* 74, 417-433.

Dorival–García, N., Junza, A., Zafra–Gómez, A., Barrón, D., Navalón, A., 2016. Simultaneous determination of quinolone and β-lactam residues in raw cow milk samples using ultrasound–assisted extraction and dispersive–SPE prior to UHPLC-MS/MS analysis. *Food Control* 60, 382 – 393.

Dudlova, A. & Juriš, P., 2014. Patogény v odpadových vodách a kaloch z čistiarní odpadových vôd. *Slovenský Veterinársky Časopis* 3-4, 118 – 122. [Pathogenes in wastewaters and sewage sludge from wastewater treatment plants. *Slovak Veterinary Journal* 3-4, 118 – 122].

European Centre for Disease Prevention and Control. ECDC, Stockholm, 2016. Summary of the latest data on antibiotic consumption in the European Union. *ESAC-Net surveillance data*, November 2016, 1-11.

European Commission. Antimicrobial Resistance, Antibiotics in the EU-Use & Prescription 2016. *Special Eurobarometer 445, Antimicrobial Resistance* 1-2. http://ec.europa.eu/commfrontoffice/publicopinion/index.cfm/General/index, http://ec.europa.eu/dgs/health_food-safety/.

Fedorova, G., Nebesky, V., Randak, T., Grabic, R., 2014. Simultaneous determination of 32 antibiotics in aquaculture products using LC-MS/MS. *Chem. Pap.* 68, 29-36.

Garrett, J. H., 2015. An Era of Antimicrobial Resistance: Imminent Threats of Superbugs to Vascular Access Clinicians and Patients. *JAVA* 20, 133-136.

Guardabassi, L., Lo Fo Wong, D. M. A., Dalsgaard, A., 2002. The effects of tertiary wastewater treatment on the prevalence of antimicrobial resistant bacteria. *Water Res.* 36, 1955-1964.

Healy, M. G., Fenton, O., Forrestal, P. J., Danaher, M., Brennan, R. B., Morrison, L., 2016. Metal concentrations in lime stabilised, thermally dried and anaerobically digested sewage sludges. *Waste Manage.* 48, 404 – 408.

Hollis, A. & Ahmed, Z., 2014. The path of least resistance: Paying for antibiotics in non-human uses. *Health Policy* 118, 264-270.

Huang, J. J., Xi, J., Hu, H. Y., Li, Y., Lu, S. Q., Tang, F., Pang, Y. CH., 2016. UV light tolerance and reactivation potential of tetracycline-resistant bacteria from secondary effluents of a wastewater treatment plant. *J. Environ. Sci. 41,* 146-153.

Chen, CH., Li, J., Chen, P., Ding, R., Zhang, P., Li, X., 2014. Occurrence of antibiotics and antibiotic resistances in soil from wastewater irrigation areas in Beijing and Tianjin, China. *Environ. Pollut.* 193, 94-101.

Kelessidis, A. & Stasinakis, A. S., 2012. Comparative study of the methods used for treatment and final disposal of sewage sludge in European countries. *Waste Manage.* 32, 1186-1195.

Khan, H. A., Ahmad, A., Mehboob, R., 2015. Nosocomial infections and their control strategies. *APJTB* 5(7), 509-514.

Lefebvre, A., Laporte, S., Faure, S., Tiv, M., Chavanet, P., Belpois-Duchamp, C., Astruc, K., Aho-Glélé, L. S., 2015. Information concerning multidrug-resistant bacterial colonization or infection in the medical transfer letter. *Med. Mal Infect.* 45, 286-292.

Mackuľak, T., Škubák, J., Grabic, R., Ryba, J., Birošová, L., Fedorova, G., Špalková, V., Bodík, I., 2014. National study of illicit drug use in Slovakia based on wastewater analysis. *Sci. Total Environ.* 494-495, 158-165.

Mackuľak, T., Mosný, M., Škubák, J., Grabic, R., Birošová, L., 2015. Fate of psychoactive compounds in wastewater treatment plant and the possibility of their degradation using aquatic plants. *Environ. Toxicol. Phar.* 39, 969-973.

Mackuľak, T., Birošová, L., Gál, M., Bodík, I., Grabic, R., Ryba, J., Škubák, J., 2016a. Wastewater analysis: the mean of the monitoring of frequently prescribed pharmaceuticals in Slovakia. *Environ. Monit. Assess.* 188(8), DOI 10.1007/s10661-015-5011-7, 1-12.

Mackuľak, T., Bodík, I., Hasan, J., Grabic, R., Golovko, O., Vojs-Staňová, A., Gál, M., Naumowicz, M., Tichý, J., Brandeburová, P., Híveš, J., 2016b. Dominant psychoactive drugs in the Central European Region: A wastewater study. *Forensic Sci. Int.* 267, 42-51.

Mackuľak, T., Marton, M., Radičová, M., Vojs-Staňová, A., Grabic, R., Bírošová, L., Nagyová, K., Vojs, M., Bodík, I., Brandeburová, P., Gál, M., 2017. Monitoring of micropollutants and resistant bacteria in wastewater and their effective removal by boron doped diamond electrode. *Monatsh. Chem.* 148, 539-548.

Magiorakos, A. P., Srinivasan, A., Carey, R. B., Carmeli, Y., Falagas, M. E., Giske, C. G., Harbarth, S., Hindler, J. F., Kahlmeter, G., Olsson-Liljequist, B., Paterson, D. L., Rice, L. B., Stelling, J., Struelens, M. J., Vatopoulos, A., Weber, J. T., Monnet, D. L., 2012. Multidrug-resistant, extensively drug-resistant and pandrug-resistant bacteria: an international expert proposal for interim standard definitions for acquired resistance. *Clin. Microbiol. Infec.* 18, 268-281.

Maulini–Duran, C., Artola, A., Font, X., Sánchez, A., 2013. A systematic study of the gaseous emissions from biosolids composting: Raw sludge versus anaerobically digested sludge. *Bioresource Technol.* 147, 43 – 51.

Mutiyar, P. K. & Mittal, A. K., 2014. Occurrence and fate of selected human antibiotics in influents and effluents of sewage treatment plant and effluent-receiving river Yamuna in Delhi (India). *Environ. Monit. Assess.* 186, 541-557.

Naidoo, S. & Olaniran, A. O., 2014. Treated Effluent Wastewater as a Source of Microbial Pollution of Surface Water Resources. *Int. J. Env. Res. Pub. He.* 11, doi: 10.3390/ijerph110100249, 249-270.

Palacios, O. A., Contreras, C. A., Muñoz-Castellanos, L. N., González-Rangel, M. O., Rubio-Arias, H., Palacios-Espinosa, A., Nevárez-Moorillón, G. V., 2017. Monitoring of indicator and multidrug resistant bacteria in agricultural soils under different irrigation patterns. *Agr. Water Manage.* 184, 19-27.

Picot-Guéraud, R., Batailler, P., Caspar, Y., Hennebique, A., Mallaret, M.-R., 2015. Bacteremia caused by multidrug-resistant bacteria in a

French university hospital center: 3 years of collection. *Am. J. Infect. Control* 43, 960-964.

Reinthaler, F. F., Feierl, G., Galler, H., Haas, D., Leitner, E., Mascher, F., Melkes, A., Posch, J., Winter, I., Zarfel, G., Marth, E., 2010. ESBL-producing *E. coli* in Austrian sewage sludge. *Water Res.* 44, 1981-1985.

Rizzo, L., Manaia, C., Merlin, C., Schwartz, T., Dagot, C., Ploy, M. C., Michael, I., Fatta-Kassinos, I., 2013. Urban wastewater treatment plants as hotspots for antibiotic resistant bacteria and genes spread into the environment: A review. *Sci. Total Environ.* 447, 345-360.

Rodriguez-Mozaz, S., Chamorro, S., Marti, E., Huerta, B., Gros, M., Sànchez-Melsió, A., Borrego, C. M., Barceló, D., Balcázar, J. L., 2015. Occurrence of antibiotics and antibiotic resistance genes in hospital and urban wastewaters and their impact on the receiving river. *Water Res.* 69, 234-242.

Song, B., Wang, G., Shoemaker, N. B., Salyers, A. A., 2009. An unexpected effect of tetracycline concentration: growth phase-associated excision of the Bacteroides mobilizable transposon NBU1. *J. Bacteriol.* 191, 1078–82.

Stiborova, H., Kolar, M., Vrkoslavova, J., Pulkrabova, J., Hajslova, J., Demnerova, K., Uhlik, O., 2017. Linking toxicity profiles to pollutants in sludge and sediments. *J. Hazard. Mater.* 321, 672-680.

Theuretzbacher, U., 2013. Global antibacterial resistance: The never-ending story. *J. Glob. Antimicrob. Re.* 1, 63-69.

Valderrama, C., Granados, R., Cortina, J. L., 2013. Stabilization of dewatered domestic sewage sludge by lime addition as raw material for the cement industry: Understanding process and reactor performance. *Chem. Eng. J.* 232, 458 – 467.

Wang, J., Mao, D., Mu, Q., Luo, Y., 2015. Fate and proliferation of typical antibiotic resistance genes in five full-scale pharmaceutical wastewater treatment plants. *Sci. Total Environ.* 526, 366-373.

Zhang, CH., Xu, X., Pu, S., Huang, S., Sun, J., Yang, S., Zhang, L., 2014. Characterization of carbapenemases, extended spectrum β-lactamases, quinolone resistance and aminoglycoside resistance determinants in carbapenem-non-susceptible Escherichia coli from a teaching hospital in Chongqing, Southwest China. *Infect. Genet. Evol.* 27, 271-276.

BIOGRAPHICAL SKETCHES

Kristina Lépesová

Affiliation: Department of Nutrition and Food Quality Assessment, Faculty of Chemical and Food Technology, Slovak University of Technology, Bratislava, Slovakia

Education: In 2014 she graduated in Nutrition and Health Protection at the Faculty of Chemical and Food Technology STU in Bratislava, Slovakia and gained Ing. (MSc.)

Professional Appointments: She works in the area of microbiology for more than 5 years. After she has completed the second level of higher education with honors, she has log on the PhD study in the field of Biochemistry. In 2015 she has received a FEMS grant (Young Scientist Meeting Grant) as well as prestigious award in the form of conference passport by BASF Ltd. at interactive conference PREVEDA. Her research is focused mainly on the issue of antibiotic resistance in food and the environment, monitoring of the natural and synthetic substances biological effects, or the study of intestinal microbiota in relation to nutrition. She is the author or co-author of several scientific articles in peer-reviewed scientific journals and she participates in various grant projects as a principal investigator or co-worker. She has also presented her results several times at domestic or international conferences.

Honors:

- FEMS grant (Young Scientist Meeting Grant)
- Conference passport by BASF Ltd.

- 2x principal investigator of STU grant for young scientists

Publications from the Last 3 Years:

1. Lépesová, K., Mackuľak, T., Bírošová, L., 2017. Effect of Wastewater Pollution on the Development and Spread of Antibiotic Resistance in Bacteria. *Chemické listy* 111 (6), 374-380.
2. Mackuľak, T., Marton, M., Radičová, M., Vojs Staňová, A., Grabic, R., Birošová, L., Nagyová, K., Vojs, M., Bodík, I., Brandeburová, P., Gál, M., 2016. Monitoring of micropollutants and resistant bacteria in wastewater and their effective removal by boron doped diamond electrode. *Monatschefte fur Chemie*, DOI 10.1007/s00706-016-1914-4.
3. Šaková, N., Olejníková, P., Kaliňáková, B., Pangallo, D., Lakatoš, B., Nagyová, K., Valachovičová, M., Birošová, L., 2015. Comparative study of adult Slovak vegetarians and meat-eaters gut microbiota. *Journal of Food and Nutrition Research* 54, ISSN 1336 – 8672, 260 – 269.
4. Mackuľak, T., Nagyová, K., Fáberová, M., Grabic, R., Koba, O., Gál, M., Birošová, L., 2015. Utilization of Fenton–like reaction for antibiotics and resistant bacteria elimination in different parts of WWTP. *Environmental Toxicology and Pharmacology* 40, 492 – 497.
5. Hybenová, E., Birošová, L., Nagyová, K., Štofirová, J., Šaková, N., Olejníková, P., Kaliňáková, B., 2014. Testing of selected probiotic properties of lactic acid bacteria isolated from vegetarians and meat-eaters faeces. *Acta Chimica Slovaca* 7, 38—43.

Tomáš Mackuľak

Affiliation: Department of Environmental Engineering, Faculty of Chemical and Food Technology, Slovak University of Technology

Education: In 2007 he graduated in Environmental Engineering at the Faculty of Chemical and Food Technology STU in Bratislava, Slovakia and gained Ing. (MSc.). In 2011 he finished my Ph.D. in Chemical Engineering at the Faculty of Chemical and Food Technology STU in Bratislava and gained scientific - academic degree PhD.

Professional Appointments: Assoc. Prof. Dr. Tomas Mackuľak is the young scientist up to 35 years. He is currently working on analysing and development of degradation processes of micropollutants such as pharmaceuticals, drugs and antibiotic resistant microorganisms in the environment.

He co-authored several publications in scientific journals, as well as lectures and contributions to scientific conferences (as a first author in 110 contributions and studies and as a co-author of the other 160 studies, including 38 CC, 5 CA, more than 260 contributions in the conference proceedings). At the age of 34 years he is also co-author of a university four textbooks and three monographies.

He is the author and investigator of three grants supported by SUT program for young researchers, investigator of three grants financially supported by Ministry of Health (Incidence of psychoactive compounds in wastewater of Slovakia).

He is a leading expert of group for monitoring and the removal of organic micropollutants from water in the Association of WWTP experts of SR. Currently, he is the first author of five unique CC publications, which are dealing with the presence of illegal drugs in Slovak wastewaters, which were realized in Slovakia. He is also co-author of publication concerning the occurrence of antibiotics and antibiotic resistant microorganisms in wastewater and sludge in Slovakia.

Despite his young age, the investigator has many experiences with the introduction of progressive and advanced technologies into practice. He has collaborated with companies such Duslo Sala, Chemical plants Nováky, Istrochem Bratislava, de Miclén Levice, Peugeot Trnava, Bratislava Volkswagen, Fagor Ederland Slovakia, s.r.o. Enviroline Ziar nad Hronom and Marius Pedersen a.s.

Assoc. Prof. Dr. Tomas Mackuľak becomes a major expert in the issue of illicit drugs, pharmaceuticals and resistant bacteria occurrence in wastewater of Slovakia during his short scientific career. Evidence of this fact is 16 studies in international scientific journals regarding the occurrence of drugs and pharmaceuticals in wastewater of Slovak Republic.

In addition, he is author of more than 80 conference contributions in national and international conferences dealing with occurrence and degradation of drugs and pharmaceuticals during the last two years

Honors:

- "STU price for significant scientific contribution for research team (2016)"
- "STU price young scientist of the year (2015)"
- At the conferences "WATER" in 2011 and 2013 in Podebrady and "Wastewater" in 2012 and 2014 principal investigator´s several studies have been awarded (2nd place in the poster section "Applied science, technology and engineering," in 2011 for post: Enzymatic treatment of real industrial water with a high content of phenolic compounds, 2nd place in the poster section "Research and development" in 2013 for contribution: Application of wood rot fungi in the degradation of drugs and hormones occurring in wastewater, 3rd place in the poster section Forum 33: Enzymatic treatment of industrial water in the WWTP PSA Peugeot Citroen Trnava in 2012; 2nd place in the main lecture section Forum 33: Fate of selected pharmaceuticals and drugs in the WWTP in 2014).

Publications from the Last 3 Years:

1. Lépesová, K., Mackuľak, T., Birošová, L.: Effect of wastewater pollution on the development and spread of antibiotic resistance in bacteria, *Chemické listy* 111(6), 374-380, 2017 (IF 0,387)

2. Brandeburova, P., Bírošová, L., Vojs, M., Kromka, A., Gál, M., Tichý, J., Híveš, J., Mackuľak, T.: The influence of selected nanomaterials on microorganisms, *Monatsh Chem.* 148(3), 525-530 2017
3. Mackuľak, T., Marton, M., Radičová, M., Vojs Staňová, A., Grabic, R., Bírošová, L., Nagyová, K., Vojs, M., Bodík, I., Brandeburová, P., Gál, M.: Monitoring of micropollutants and resistant bacteria in wastewater and their effective removal by boron doped diamond electrode, *Monatsh Chem.* 148(3), 539-548, 2017
4. Mackuľak, T., Birošová, L., Gál, M., Bodík, I., Grabic, R., Ryba, J., Škubák, J.: Wastewater analysis: the mean of the monitoring of frequently prescribed pharmaceuticals in Slovakia, *Environmental Monitoring and Assessment* 188 (1), 1 -12, 2016
5. Mackuľak, T., Vois, M., Grabic, R., Golovko, O., Vojs Staňová, A., Birošová, L., Medveďová, B., Híveš, J., Gál, M., Kromka, A., Hanusová, A.: Occurrence of pharmaceuticals, illicit drugs, and resistant types of bacteria in hospital effluent and their effective degradation by boron-doped diamond electrodes, *Monatsh Chem.*147(1),. 97-103, 2016
6. Mackuľak, T., Birošová, L., Bodík, I., Grabic, R., Takáčová, A., Smolinská, M., Hanusová, A., Híveš, J., Gál, M.: Zerovalent iron and iron (VI): effective means for the removal of psychoactive pharmaceuticals and illicit drugs from wastewaters, *Sci. Tot. Env.* 539, 18328, 420-426, 2016
7. Mackuľak, T., Grabic, R., Gál, M., Gál, M., Birošová, L., Bodík, I.: Evaluation of different smoking habits during music festivals through wastewater analysis, *Env. Tox. Pharm.*40 (3), pp. 1015-1020, 2015
8. Mackuľak, T., Nagyova K., Grabic, R., Koba, O., Gál, M., Birošová, L.: Utilization of Fenton-like reaction for antibiotics and resistant bacteria elimination in different parts of WWTP, *Env. Tox. Pharm.* 40(2), pp. 492-497, 2015
9. Mackuľak, T., Birošová, L., Grabic, R., Škubák, J., Bodík, I: National monitoring of nicotine use in Czech and Slovak Republic based on

wastewater analysis, *Env. Sci. Pol. Res.* 22(18), pp. 14000-14006, 2015

10. Mackuľak, T., Mosný, M., Škubák, J., Grabic, R., Birošová, L.: Fate of psychoactive compounds in wastewater treatment plant and the possibility of their degradation using aquatic plants, *Env. Tox. Pharm.* 39(2), 969-973, 2015
11. Mackuľak, T., Mosný, M., Grabic, R., Golovko, O., Koba, O., Birošová, L.: Fenton-like reaction: A possible way to efficiently remove illicit drugs and pharmaceuticals from wastewater, *Env. Tox. Pharm.* 39 (2), 483-488, 2015
12. Mackuľak, T., Škubák, J., Grabic, R., Ryba, J., Birošová, L., Fedorova, G., Špalková, V., Bodík, I.: National study of illicit drug use in Slovakia based on wastewater analysis, *Sci. Tot. Env.* 494-495, 158-165, 2014
13. Birošová L., Mackuľak, T., Bodík, I., Ryba, J., Škubák, J., Grabic, R.: Pilot study of seasonal occurrence and distribution of antibiotics and drug resistant bacteria in wastewater treatment plants in Slovakia, *Science of the Total Environment* 490, 440-444, 2014
14. Špalková, Viera, Lazor, Michal, Smolinská, Miroslava, Takáčová, Alžbeta, Hutňan, Miroslav, Bodík, Igor, Ryba, Jozef, Gál, Miroslav, Žemlička, Lukáš, Pangallo, Domenico, Mackuľak, Tomáš. Enhanced hydrogen bioproduction from bird-cherry leaves using enzyme mixture. *Monatshefte für Chemie, Chemical Monthly.* 147, 1 (2016), 201-206.
15. Stanko, Štefan, Hrudka, Jaroslav, Škultétyová, Ivona, Holubec, Michal, Galbová, Kristína, Gregušová, Veronika, Mackuľak, Tomáš. CFD analysis of experimental adjustments on wastewater treatment sedimentation tank inflow zone. *Monatshefte für Chemie - Chemical Monthly [elektronický zdroj] : selected papers from the Conference XXXVIth Modern Electrochemical Methods, Czech Republic, 2016.* 148, 3 (2017), online, 585-591.
16. Stanko, Štefan, Hrudka, Jaroslav, Škultétyová, Ivona, Holubec, Michal, Galbová, Kristína, Mackuľak, Tomáš. Analysis of seasonal temperature influence on hydraulic processes in a waste water

sedimentation tank. *Monatshefte für Chemie - Chemical Monthly [elektronický zdroj] : selected papers from the Conference XXXVIth Modern Electrochemical Methods, Czech Republic, 2016.* 148, 3 (2017), online, 489-495.

17. Takáčová, Alžbeta, Smolinská, Miroslava, Ryba, Jozef, Mackuľak, Tomáš, Jokrllová, Jana, Hronec, Pavol, Čík, Gabriel. Biodegradation of Benzo[a]Pyrene through the use of algae. *Central European Journal of Chemistry*. 12, 11 (2014), 1133-1143.
18. Tichý, Jozef, Takáčová, Alžbeta, Smolinská, Miroslava, Bodík, Igor, Gál, Miroslav, Czölderová, Marianna, STANKO, Štefan, Holubec, Michal, Hrudka, Jaroslav, Vojs Staňová, Andrea, Mackuľak, Tomáš. Treatment of hazardous leachate from landfill using ultrasound/H2O2 system. *Monatshefte für Chemie - Chemical Monthly [elektronický zdroj]*. 148, 3 (2017), s. 563-570.
19. Šeršeň, František, Pavlíková, Stanislava, Jesenák, Karol, Jokrllová, Jana, Takáčová, Alžbeta, Mackuľak, Tomáš, Čík, Gabriel. Use of the modified montmorillonites for phopocatalytic degradation of 4-chlorophenol. *Fresenius Environmental Bulletin*. 24, 6B (2015), 2057-2062.
20. Mackuľak, Tomáš, Bodík, Igor, Hasan, Jamal, Grabic, Roman, Golovko, Oksana, Vojs Staňová, Andrea, Gál, Miroslav, Naumowicz, Monika, Tichý, Jozef, Brandeburová, Paula, HÍVEŠ, Ján. Dominant psychoactive drugs in the Central European region: A wastewater study. *Forensic Science International*. 267, (2016), 42-51.
21. Mackuľak, Tomáš, Bodík, Igor, Smolinská, Miroslava, Takáčová, Alžbeta, Drtil, Miloslav, Gál, Miroslav - Fáberová, Milota. Automotive industry wastewater treatment by mixture of enzymes. *Monatshefte für Chemie - Chemical Monthly*. 147, 1 (2016), 159-164.
22. Mackuľak, Tomáš, Vojs Staňová, Andrea, Gál, Miroslav, Híveš, Ján, Grabic, Roman, Tichý, Jozef. Determination of illicit drugs and their metabolites contamination on banknotes. *Monatshefte für Chemie - Chemical Monthly*. 147, 1 (2016), 39-43.
23. Mackuľak, Tomáš, Takáčová, Alžbeta, Gál, Miroslav, Marton, Marián, Ryba, Jozef. PVC degradation by Fenton reaction and

biological decomposition. *Polymer Degradation and Stability*. 120, (2015), 226-231.

24. Mackuľak, Tomáš, Škubák, Jaroslav, Grabic, Roman, Drtil, Miloslav, Bodík, Igor. Comparison of illicit drug use in three selected towns in Slovakia by wastewater analysis. In *Urban Water Journal*. 12, 7 (2015), 519-524.
25. Ivanová, Lucia, Fáberová, Milota, Mackuľak, Tomáš, Grabic, Roman, Bodík, Igor. Estimation of amount of selected pharmaceuticals sorbed onto digested sludge from wastewater treatment plant Bratislava-Petržalka. *Environmental research*. 155, (2017), 31-35.
26. Fáberová, Milota, Bodík, Igor, Ivanová, Lucia, Grabic, Roman, Mackuľak, Tomáš. Frequency and use of pharmaceuticals in selected Slovakian town via wastewater analysis. *Monatshefte für Chemie - Chemical Monthly* 148, 3 (2017), s. 441-448.
27. Čerňanský, Slavomír, Peťková, Katarína, Feketeová, Zuzana, Jurkovič, Ľubomír, Šimonovičová, Alexandra, Faragó, Tomáš, Mackuľak, Tomáš. Arsenic ashy soils in Central Slovakia and their chemical and microbiological properties. *Monatshefte für Chemie - Chemical Monthly* 148, 3 (2017), 593-600.
28. Bodík, Igor, Mackuľak, Tomáš, Fáberová, Milota, Ivanová, Lucia. Occurrence of illicit drugs and selected pharmaceuticals in Slovak municipal wastewater. *Environmental science and pollution research*. 23, 20 (2016), 21098-21105.

In: Nutrients, Wastewater and Leachate
Editor: Amimul Ahsan
ISBN: 978-1-53613-949-5

Chapter 4

ANTIBIOTIC RESISTANT COLIFORMS: FROM HUMAN GUT TO WASTEWATER

Lucia Bírošová**, Karolína Kislíková and Kristína Lépesová
Department of Nutrition and Food Quality Assessment,
Faculty of Chemical and Food Technology,
Slovak University of Technology, Bratislava, Slovakia

ABSTRACT

Gut microbiota represents a complex organ which affects human health. On the other hand, there are many factors such age, diet, stress, antibiotic therapy which influences composition and activity of gut microbiota. Human gut microbiota is composed of broad spectrum of bacterial species. Recent studies showed that this bacterial community comprise resistance genes harboured by major pathogens. These genes can be exchanged among gut bacteria, but also might interact with bacteria passing through gastrointestinal tract. Susceptible and antibiotic

* Corresponding Author Email: lucyja.birosova@gmail.com.

resistant bacteria together with resistance genes can get to wastewater with stool. In this point of view, wastewater represents another hotspot of bacterial resistance. Much more alarming issue is the presence of pharmaceuticals in subinhibitory concentrations which contributes to the development of mutations leading to antibiotic resistance. These micropolutants can also accelerate horizontal gene transfer. In this study, we have monitored prevalence of antibiotic resistant coliforms in stool of healthy adult people as well as in wastewater in the capital city of Slovakia. Total coliforms in gut microbiota from probands ranged from 3 to 9 log CFU/g of stool. Majority samples contained ampicillin and tetracycline resistant coliforms, however many species pose intrinsic resistance to these antibiotics. In case of *Escherichia coli*, we have detected tetracycline resistance in almost half samples. We have also studied influent wastewater of two wastewater treatment plants, which are situated in the same city like probands. Total coliforms ranged from 4.5 to 4.8 log CFU/mL and majority was resistant to ampicillin and gentamicin. Beside these mentioned antibiotics we have also monitored resistance to chloramphenicol and tetracycline. Coliforms resistant to all studied antibiotics were detected in influent wastewater. Obtained data show, that antibiotic resistance is a global problem, concerning also healthy individuals and can be disseminated to the environment through wastewater.

Keywords: microbiota, coliform bacteria, antibiotic resistance, wastewater

INTRODUCTION

The intestinal microbiota is composed from a high variety of microorganisms that consists of over 800 different bacterial species (Ley et al., 2006). Microorganisms are introduced into the body gradually from childhood and colonization of gastrointestinal tract (GIT) occurs immediately after the birth. While the infant microbiota is relatively simple, it becomes more complex with increasing age (Blaut and Clavel, 2007). The microbiota is also important in terms of producing some vitamins (B12 and K) and enzymes that promote proper food digestion. Commensal microbiota also produces acids and bacteriocins that inhibit the growth and adherence of harmful microorganisms to the intestinal epithelium. It is necessary to retain the correct representation of the intestinal microbiota, in order to properly secure the immune and digestive function of the intestine (Mařatka, 1988). Composition of the intestinal microbiota is affected by many factors such age, physiology, diet, obesity, inflammatory diseases, chronic diseases, but in large extent by antibiotic treatment. Introduction of antibiotics into medical praxis saved a many lives. Although they fight against pathogenic bacteria that are responsible for various infectious diseases, antibiotics can also disrupt the overall intestinal microbial community. This may be reflected in dysmicrobia, post-antibiotic diarrhoea or colitis (Zbořil et al., 2005; Rolain, 2013). Antibiotic treatment reduces the variety of intestinal microbiota in both children and adults. However intestinal microbiota in humans differs. Therefore antibiotic treatment does not have to affect everyone (Penders et al., 2006).

Nowadays, the effect of antibiotics on the intestinal microbiota is intensively studied. It is well known, that subinhibitory concentrations of antimicrobial agents can lead to the development of antibiotic resistance. This could also happen, when people misuse antibiotics during their treatment. Antibiotic resistant bacteria can also get to gastrointestinal tract through food. So it is not surprising that the human intestinal microbiota is a reservoir of antibiotic resistance genes that can be transmitted from one bacterium to the other. Genes conferring to antibiotic resistance are present among the gut microbiota of humans and most of these genes are harboured by strictly anaerobic gut commensals (van Schaik, 2015). Analysis of 252 faecal samples from different countries revealed the presence resistance genes for 50 antibiotics from different classes and subclasses, at an average of 21 antibiotic resistance genes per sample. These samples were collected from Spain, Italy, Denmark, France, Malawi, USA and Japan (Forslund et al., 2013). Hu et al. (2013) published similar data from 162 individuals in China, Denmark and Spain. They have identified 1093 antibiotic resistance genes, which were overlapping with that analysed by previously mentioned research. Other research has confirmed that individuals from countries, where antibiotic consumption is not so high, have lower levels of antibiotic resistance genes than those in countries where ATB use is significantly more frequent. The use of antibiotics in human and veterinary medicine therefore has a significant impact on the amount of ATB resistance genes in the intestinal microbiota (van Schaik, 2015).

The large majority of the bacteria that populate the human gut are strictly anaerobic. Two phyla, the *Bacteroidetes* and *Firmicutes*, commonly dominate the gut microbiota of healthy adults. Beside strict anaerobes, gut microbiota is colonised also by several facultative anaerobic bacteria from the family *Enterobacteriaceae*. Nowadays, these ubiquitous representatives of gut microbiota are of interest because some groups such *Escherichia coli* or *Klebsiella pneumoniae* emerges as multidrug resistant nosocomial pathogens of major importance in the last few decades (van Schaik, 2015). A major problem represents extended spectrum beta-lactamase (ESBL) producing strains that are capable to degrade penicillins, cephalosporins, but not carbapenems or cefamycins. Hospitals, healthcare facilities, retirement homes as well as healthy animals are hot spots of these strains and genes. Their risk lies precisely in the spread of resistance via plasmid-encoded beta-lactamase (Giske, 2013). Other problem represents AmpC beta-lactamases which are able to cleave cephamycins, but show sensitivity to fourth-generation carbapenems and cephalosporins. Genes encoding AmpC enzymes are located on plasmids or chromosomes. Representatives of this type of resistance include *Citrobacter freundii*, *E. coli*, *Enterobacter* spp., *Serratia marcescens*, *Yersinia enterocolitica*, *Klebsiella* spp. or *Salmonella* spp. (Bush, 1989). WHO list of priority resistant pathogens also include carbapenem resistant *Enterobacteriaceae*, which are of a critical priority (WHO, 2017). These bacteria produce carbapenemases belonging to the beta-lactamase group. They are able to hydrolyse cephalosporin, penicillin and carbapenem. Carbapenemases are mostly encoded on plasmids. Bacterial carbapenemase producers often exhibit resistance to broad spectrum cephalosporin (Queenan, 2007; Giske, 2013).

Majority of antibiotic resistant bacteria and antibiotic resistance genes end in municipal sewerage in faeces. Municipal wastewater flows into wastewater treatment plants (WWTP), which clean water from different pollutants. Whereas many recent studies showed, that cleansing process is often not very effective in elimination of resistance genes and bacteria (Mackuľak et al., 2017; Mackuľak et al., 2015a). Beside antibiotic resistant bacteria, wastewater contains lot of different drugs and pharmaceuticals in very low concentrations, which can cause mutations leading to the development of antibiotic resistance and increase horizontal gene transfer. Both mechanisms lead to development and further spread of antibiotic resistance. According to this fact WWTP represents hotspot of the antibiotic resistance.

The aim of this study was the monitoring of antibiotic resistant coliform bacteria in faeces of healthy adults as well as in influent and effluent wastewater in Slovakia.

MATERIAL AND METHODS

Characterization of Stool Samples

Stool samples come from healthy human subjects recruited in Bratislava (Slovakia) and its surrounding area. All 20 volunteers (10 male and 10 female) were born and raised in Slovakia. 10 volunteers with common diet (meat-eaters). Group of vegetarians consisted of 5 lacto-ovo-vegetarians and 5 semi-vegetarians with the distribution of BMI >16. Age of all subjects ranged from 20 to 40 years (Table 1). Lacto-ovo-vegetarians did not consume meat in any form, but ate animal products such as milk, cheese and eggs. A vegan diet was considered for those who did not consume animal products. Semi-vegetarians have stated that they consume animal products and also poultry and fish meat. The omnivorous consumed all type of meat (pork, red meat, poultry and fish) minimal 5 times a week. All volunteers were healthy, and had not undergone antibiotic or other medical therapy in last 6 months.

Location and Characterization of WWTPs

Samples of influent and effluent wastewater as well as stabilized sludge were collected from two WWTPs in Slovakia (Bratislava, Petržalka) situated in the capital city. Bratislava is the most populated city in the Slovak Republic, a cosmopolitan capital with approximately 466,000 inhabitants. Petržalka is considered to be the largest suburb of Bratislava, with the highest population density in Slovakia (125,000 inhabitants) as well as all of Central Europe. WWTPs are receiving wastewater originated in Bratislava, while the wider central part of the city sewerages to the Central Bratislava WWTP. The sewerage leading the effluent to this WWTP is approximately 30 km long. Whereas Petržalka WWTP treats wastewater from the largest housing estate in Bratislava (the length of sewerage is 10 km) (Mackuľak et al., 2016). Exact characterization of studied plants was referred in Mackuľak et al. (2014) and Mackuľak et al. (2015b).

Sampling Methodology

Fecal samples were collected from April to May in sterile falcon tubes and lyophilized and stored at - 40°C. Wastewater sampling points were situated on the influent and effluent stream of WWTP. For sample collection during 24 hours in 15 minutes intervals an automatic sampler device was used, while stabilized sludge was taken directly from a chamber filter press. Influent and effluent wastewaters were immediately transported in sterile falcon tubes to the laboratory for microbiological analysis.

Detection of Antibiotic Resistant Coliform Bacteria

In the first step, 0.1 g of lyophilized stool sample resuspended in 10 mL of physiological saline. Subsequently, serial 10-fold dilutions of stool as well as influent wastewater were prepared and 0.1 mL aliquots were applied on antibiotic and antibiotic-free diagnostic media (Chromocult Coliform Agar-Merck, Germany). In case of the effluent wastewater, the volume of 10 mL was filtered through GH polypro membrane (0.2 μm, Pall Corporation, USA) and placed on selective media supplemented with antibiotics. For resistance detection assessment different antibiotics (ampicillin, gentamicin, ciprofloxacin, chloramphenicol and tetracycline) in concentrations defining resistance breakpoints given by Clinical Laboratory Standards Institute (CLSI) were used. Colonies of total and resistant coliform bacteria were counted manually after cultivation for 24 hours at 37°C. Each experiment was run in parallel and repeated three times.

RESULTS AND DISCUSSION

Occurrence of Resistant Coliform Bacteria in Human Gut

The human gut contains a densely populated microbial ecosystem, which offers ample opportunities for the horizontal transfer of genetic material, including antibiotic resistance genes. The horizontal transfer of genetic material, including antibiotic resistance genes, through conjugation and transduction is a frequent event in the gut microbiota, but mostly involves non-pathogenic gut commensals as these dominate the microbiota of healthy individuals (van Schaik, 2015). According to this fact, we have studied the prevalence of antibiotic resistant coliforms in 20 stool samples of healthy people in Slovakia (Table 1).

Total coliform bacteria ranged from 3.20 to 9.49 log CFU/0.1 g of dry lyophilized stool sample. Resistance against antibiotics of different classes was studied. ß-Lactam (ampicillin) was chosen, because of its frequent prescription in Slovakia (Birosova et al., 2014). Concentration of antibiotics was selected according to the resistance breakpoints given by Clinical and Laboratory Standards Institute (CLSI).

Table 1. Characterization of probands; representation of total and proportion of resistant coliforms in stool samples

Sex	Age	Diet	Total CFB (log CFU/0.1g)	AMP	GEN	CIP	CHL	TET
male	22	LV	7.08	0.66	0.61	0.68	0.58	0.60
female	27	LV	5.75	ND	ND	0.74	ND	0.67
female	26	SV	3.20	ND	ND	ND	ND	0.68
male	27	LV	5.45	0.77	ND	0.67	ND	0.71
male	29	SV	4.54	0.94	ND	ND	ND	0.85
female	31	LV	4.41	0.95	0.78	ND	0.68	0.89
female	32	SV	9.20	0.62	0.49	0.41	0.41	0.43
female	37	SV	3.72	0.81	ND	ND	ND	0.97
male	34	SV	5.40	0.74	0.55	0.61	0.61	0.66
male	33	LV	3.22	ND	ND	ND	ND	ND
female	23	ME	3.32	ND	ND	ND	ND	0.82
female	26	ME	4.57	ND	0.44	0.52	ND	0.57
female	28	ME	4.14	ND	ND	ND	0.66	0.65
male	30	ME	4.39	0.74	0.38	ND	0.66	0.64
male	25	ME	4.64	0.80	ND	ND	0.65	0.89
male	37	ME	9.49	0.76	0.49	0.47	ND	0.53
female	38	ME	6.15	0.85	0.66	0.71	0.72	0.80
male	31	ME	5.57	0.91	0.87	0.77	ND	0.90
female	32	ME	3.99	ND	ND	ND	ND	0.72
male	36	ME	4.20	ND	ND	ND	ND	0.55

Key: LV – lacto-ovo-vegetarian, SV-semi vegetarian, ME – meat-eater, CFB – coliform bacteria, CFU – colony formatting unit, AMP – ampicillin, GEN – gentamicin, CIP – ciprofloxacin, CHL – chloramphenicol, TET – tetracycline, ND – not detected

If bacteria are able to grow in the presence of such high antibiotic concentrations, they are considered to be resistant. Table 1 shows data about the prevalence of antibiotic resistant coliform bacteria in stool. In case of each antibiotic, resistance is expressed as a proportion of antibiotic resistant bacteria number to total count. Ampicillin resistant coliforms were detected in 12 samples. It is not surprising, however majority of coliforms display intrinsic resistance to this antibiotic. Gentamicin and ciprofloxacin resistance was observed in 9 samples. Nowadays the use of chloramphenicol in Slovakia is delimited and it is prescribed only in very urgent cases. Chloramphenicol resistance was registered in 8 samples. It can be assumed that this resistance is the result of cross resistance caused by similar resistance mechanisms, probably the efflux pumps overproduction (Birosova et al., 2014). Almost all stools contained tetracycline resistant coliform bacteria. Genes providing resistance to the antibiotic tetracycline (tet32, tet40, tetO, tetQ and tetW) are present in the microbiota of all individuals and are also the most abundant family of resistance genes (Forslund et al., 2013; Hu et al., 2013). Similarly to ampicillin, some coliforms dispose of intrinsic resistance to tetracycline. If we compare abundance of resistants according to diet, number of positive samples is co-equal in both groups.

Table 2. Characterization of probands; representation of total and proportion of resistant *E. coli* in stool samples

Sex	Age	Diet	Total *E. coli* (log CFU/0.1g)	AMP	GEN	CIP	CHL	TET
man	22	LV	3.17	ND	ND	ND	ND	ND
woman	27	LV	4.45	ND	ND	ND	ND	0.86
woman	26	SV	ND	ND	ND	ND	ND	ND
man	27	LV	4.15	ND	ND	ND	ND	0.72
man	29	SV	ND	ND	ND	ND	ND	ND
woman	31	LV	ND	ND	ND	ND	ND	ND
woman	32	SV	ND	ND	ND	ND	ND	ND
woman	37	SV	3.30	0.90	ND	ND	ND	0.92
man	34	SV	4.68	ND	ND	ND	ND	0.65
man	33	LV	ND	ND	ND	ND	ND	ND
woman	23	ME	ND	ND	ND	ND	ND	ND
woman	26	ME	3.56	ND	0.56	0.67	ND	0.73
woman	28	ME	3.84	ND	ND	ND	0.52	0.60
man	30	ME	3.49	ND	0.49	ND	0.77	0.73
man	25	ME	ND	ND	ND	ND	ND	ND
man	37	ME	8.90	ND	ND	ND	ND	0.57
woman	38	ME	4.30	ND	ND	ND	ND	0.80
man	31	ME	ND	ND	ND	ND	ND	ND
woman	32	ME	ND	ND	ND	ND	ND	ND
man	36	ME	ND	ND	ND	ND	ND	ND

Key: LV – lacto-ovo-vegetarian, SV-semi vegetarian, ME – meat-eater, CFU – colony formatting unit, AMP – ampicillin, GEN – gentamicin, CIP – ciprofloxacin, CHL – chloramphenicol, TET – tetracycline, ND – not detected.

No antibiotic resistant coliform bacteria were detected only in stool sample of 33 old lacto-ovo-vegetarian man. On the other hand, resistance to all applied antibiotics was observed in 3 vegetarians (2 men, 1 woman) and 1 omnivore man. Multidrug resistance (MDR) is defined as resistance to 3 or more antibiotic classes (Magiorakos et al., 2012). If we exclude ampicillin and tetracycline, we have 3 stool samples with proportion of resistance higher than 0.5. Iit is possible that these samples contain MDR coliforms.

Table 2 shows abundance of antibiotic resistant *E. coli* in studied samples of human gut microbiota. *E. coli* was detected only in 10 samples and ranged from 3.17 to 8.90 log CFU/0.1 g of dry lyophilized stool sample. Ampicillin and ciprofloxacin resistance was observed only in one sample. Gentamicin and chloramphenicol resistant *E. coli* was also registered in 2 samples. Similarly like in case of coliforms, the most samples contained tetracycline resistant *E. coli*.

Presence of multidrug resistant *E. coli* is probable in sample from 26 year old meat-eater woman. Studies have shown that anaerobic gut commensals form the main reservoir of antibiotic resistance genes. These genes can circulate among the gut microbiota and can be transferred from mostly harmless gut commensals to opportunistic pathogens (van Schaik, 2015).

Occurrence of Resistant Coliform Bacteria in Wastewater

Majority of our microbiome as well as resistome in faeces ends in sewerage system and consequently in WWTP. However probands included in our study live or work in capital city of Slovakia (Bratislava), we have monitored the influent and effluent wastewater from two biggest WWTP in Bratislava. Figure 1 shows occurrence of total and resistant coliforms in influent and effluent from the biggest WWTP in Slovakia, which treats wastewater for majority of the city. Number of total coliforms at the influent was 4.53 log CFU/mL and after WWTP treatment decreased to 1.86 log CFU/mL of effluent water. Majority of these coliforms were detected as *E. coli* (Figure 2). As it was mentioned before, many coliform bacteria pose intrinsic resistance to ampicillin, so it is not surprising, that more than 80% of coliforms in influent wastewater were resistant to this antibiotic. On the other hand, about 70% of *E. coli* in influent wastewater was also ampicillin resistant. This phenomenon reflects frequent prescription of beta-lactam antibiotics in our country, as well as presence of beta-lactamases (Birosova et al., 2014). We have also observed high prevalence of gentamicin, ciprofloxacin and tetracycline resistance of coliform bacteria in the influent. Higher occurrence of tetracycline resistant strains also correlates with its high prevalence in gut microbiota, as we have shown previously (Table 1). In case of influent *E. coli*, tetracycline resistant strains have the lowest representation from all studied resistant strains. At the effluent water, resistance of coliform bacteria to all applied antibiotics was observed. *E. coli* in the effluent from the biggest WWTP in Slovakia was resistant only to fluoroquinolone antibiotic ciprofloxacin. This could be connected with frequent use and prescription of ciprofloxacin for treatment of urinary infection, while majority is caused by *E. coli*.

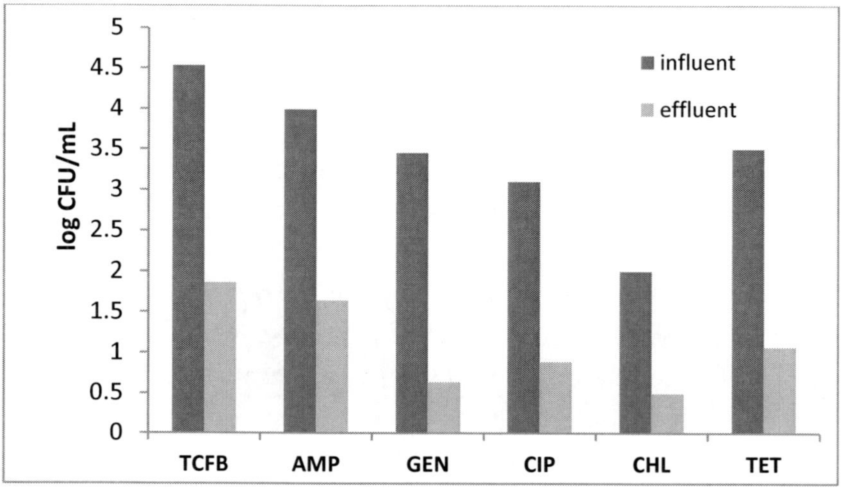

TCFB – total coliform bacteria, CFU – colony formatting unit, AMP – ampicillin, GEN – gentamicin, CIP – ciprofloxacin, CHL – chloramphenicol, TET – tetracycline.

Figure 1. Occurrence of total and resistant coliform bacteria in WWTP Bratislava influent and effluent.

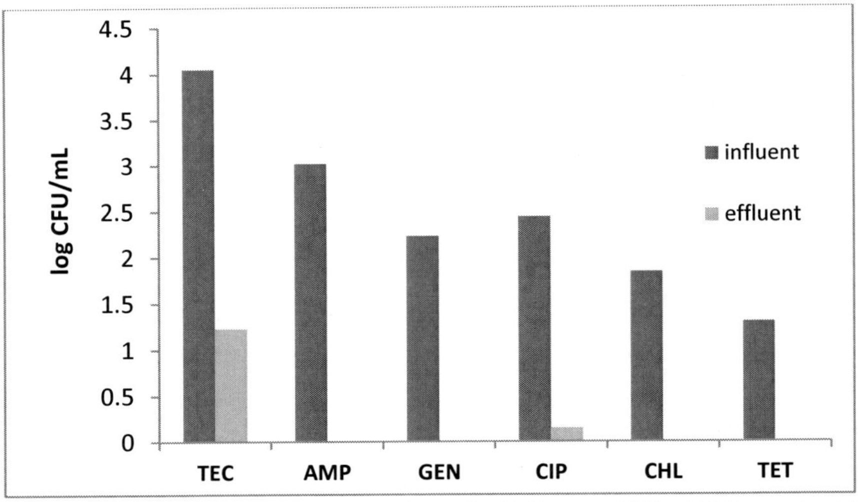

TEC – total *E. coli*, CFU – colony formatting unit, AMP – ampicillin, GEN – gentamicin, CIP – ciprofloxacin, CHL – chloramphenicol, TET – tetracycline.

Figure 2. Occurrence of total and resistant *E. coli* in WWTP Bratislava influent and effluent.

Figure 3 shows prevalence of coliform bacteria in wastewater at the influent and effluent from WWTP Petržalka, which treats wastewater from the largest housing estate in Bratislava and some small border Austrian villages. Number of coliform bacteria in influent wastewater (3.65 log CFU/mL) was smaller compared to WWTP Bratislava and 60% were represented by *E. coli*. Similarly to first monitored WWTP, we have observed coliforms resistant to all applied antibiotics, while chloramphenicol resistance was represented in lower range. At the effluent water we have detected only ampicillin and gentamicin resistant strains. Chloramphenicol and tetracycline resistant *E. coli* was not detected in influent. Number of *E. coli* at the effluent water was under the limit of detection (Figure 4). Both WWTPs have the same treating process. They use a mechanical and biological stage (stage nitrification only); the produced sludge is digested, and produced biogas energy is recovered (Mackuľak et al., 2014). Lower and different representation of total and resistant bacteria in effluent water can be caused by their entrapping in sewage sludge or in sewerage biofilm (Birosova et al., 2014).

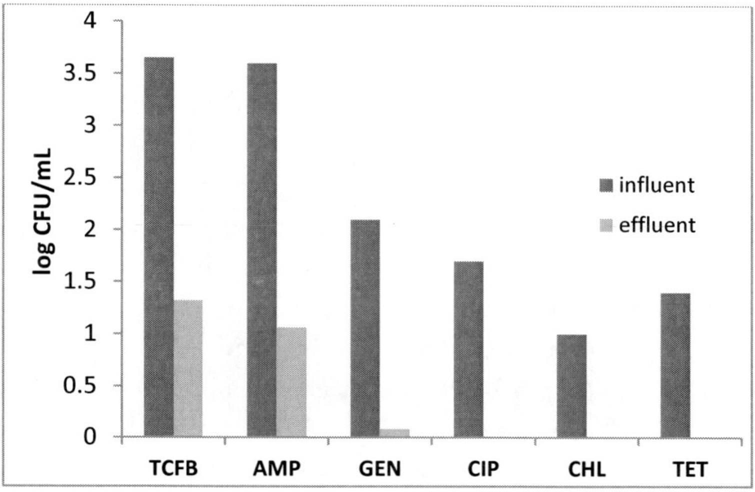

TCFB – total coliform bacteria, CFU – colony formatting unit, AMP – ampicillin, GEN – gentamicin, CIP – ciprofloxacin, CHL – chloramphenicol, TET – tetracycline.

Figure 3. Occurrence of total and resistant coliform bacteria in WWTP Petržalka influent and effluent.

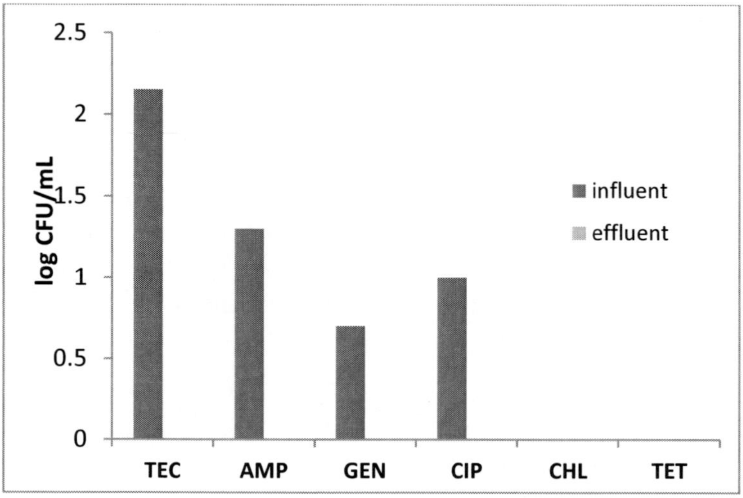

TEC – total *E. coli*, CFU – colony formatting unit, AMP – ampicillin, GEN – gentamicin, CIP – ciprofloxacin, CHL – chloramphenicol, TET – tetracycline.

Figure 4. Occurrence of total and resistant *E. coli* in WWTP Petržalka influent and effluent.

Conclusion

Antibiotic resistance is an emerging problem which threatens the effectivity of recent human and animal treatment. Recently many studies show, that resistance genes and resistant bacteria are not occurring only in hospitals or environment but also in gut microbiota of healthy individuals. This work confirms these findings. Our results show that microbiota of healthy individuals, no matter the age, sex or diet, contain different antibiotic resistant coliform bacteria. We have also confirmed, that tetracycline resistance is very common and disseminated in gut microbiota, whereas 95% of stool samples contained tetracycline resistant coliform bacteria. Our results also showed that wastewater treatment plants represent hot spots of antibiotic resistance. Influent wastewater contains high number of different antibiotic resistant strains of coliform bacteria. On the other hand, wastewater treatment plant is not totally effective in elimination of microorganisms and antibiotic resistant bacteria are also presented in treated water which ends in the environment. According to this fact, wastewater treatment plants contribute to the development of antibiotic resistance but also to its dissemination to the environment.

Acknowledgments

This study was created by realization of research project: "Health effects of plant food and the possibility of reduction of health risks", ITMS code: 26240220022 and by the Scientific Grant Agency VEGA (Grant number VEGA 1/0096/17), by the Slovak Research and Development Agency (Grant number APVV-16-0171).

References

Birošová, L., Mackuľak, T., Bodík, I., Ryba, J., Škubák, J., Grabic, R., 2014. Pilot study of seasonal occurrence and distribution of antibiotics and drug resistant bacteria in wastewater treatment plants in Slovakia. *Sci. Total Environ.* 490, 440-444.

Blaut, M., Clavel, T., 2007. Metabolic diversity of the intestinal microbiota: implications for health and disease. *J. Nutr.* 137, 751S-755S.

Bush, K., 1989. Characterization of beta-lactamases. *Antimicrob. Agents Chemother.* 33(3), 259-263.

Forslund, K., Sunagawa, S., Kultima, J. R., Mende, D. R., Arumugam, M., Typas, A., Bork, P., 2013. Country-specific antibiotic use practices impact the human gut resistome. *Genome Res.* 23, 1163–1169.

Giske, Ch. G., Martinez, L. M., Cantón, R., Stefani, S., Skov, R., Glupszynski, Y., Nordmann, P., Wooton, M., Miriagou, V., Simonsen, G. S., Zemlickova, H., Cohen-Stuart, J., Gniadkowski, M., 2013. EUCAST guidelines for detection of resistance mechanisms and specific resistance of clinical and/or epidemiological importance. *European committee on antimicrobial susceptibility testing.* Version 1.0. December 2013.

Hu, Y, Zang, X., Quin, J., Lu, N., Cheng, G., Wu, N., Pan, Y, Li, J., Zhu, L., Wang, X., Meng, Z., Zhao, F., Liu, D., Ma, J., Qin, N., Xiang, Ch., Xiao, Y., Li, L., Yang, H., Wang, J., Yang, R., Gao, G.F., Wang, J., Zhu, B., 2013. Metagenome-wide analysis of antibiotic resistance genes in a large cohort of human gut microbiota. *Nat. Commun.* 4, 2151.

Ley, E. R., Peterson, D. A., Gordon, J. I., 2006. Ecological and evolutionary forces shaping microbial diversity in the human intestine. *Cell* 124(4), 837 – 848.

Mackuľak, T., Škubák, J., Grabic, R., Ryba, J., Birošová, L., Fedorova, G., Špalková, V., Bodík, I., 2014. National study of illicit drug use in Slovakia based on wastewater analysis. *Sci. Total Environ.* 494-495, 158-165.

Mackuľak, T., Nagyova K., Grabic, R., Koba, O., Gál, M., Birošová, L. 2015a. Utilization of Fenton-like reaction for antibiotics and resistant bacteria elimination in different parts of WWTP, *Env. Tox. Pharm.* 40(2), 492-497.

Mackuľak, T., Mosný, M., Škubák, J., Grabic, R., Birošová, L., 2015b. Fate of psychoactive compounds in wastewater treatment plant and the possibility of their degradation using aquatic plants. *Environ. Toxicol. Pharmacol.* 39, 969-973.

Mackuľak, T., Birošová, L., Gál, M., Bodík, I., Grabic, R., Ryba, J., Škubák, J., 2016. Wastewater analysis: the mean of the monitoring of frequently prescribed pharmaceuticals in Slovakia. *Environ. Monit. Assess.* 188(8), 1-12.

Mackuľak, T., Marton, M., Radičová, M., Vojs Staňová, A., Grabic, R., Bírošová, L., Nagyová, K., Vojs, M., Bodík, I., Brandeburová, P., Gál, M. 2017. Monitoring of micropollutants and resistant bacteria in wastewater and their effective removal by boron doped diamond electrode, *Monatsh Chem.* 148(3), 539-548.

Magiorakos, A-P., Srinivasan, A., Carey, R. B., Carmeli, Y., Falagas, M. E., Giske, C. G., Harbarth, S., Hindler, J. F., Kahlmeter, G., Olsson-Liljequist, B., Paterson, D. L., Rice, L. B., Stelling, J., Struelens, M. J., Vatopoulos, A., Weber, J. T., Monnet, D. L., 2012. Multidrug-resistant, extensively drug-resistant and pandrug-resistant bacteria: an international expert proposal for interim standard definitions for acquired resistance. *Clin. Microbiol. Infect.* 18, 268-281.

Mařatka, Z., 1988. *Clinical gastroenterology (in Czech) Klinická gastroenterologie.* ISBN 80-718-4561-2, Praha: Avicenum, Czech Republic, 322-329.

Penders, J., Thijs, C., Vink, C., Stelma, F. F., Snijders, B., Kummeling, I., van den Brandt, P. A., Stobberingh, E. E., 2006. Factors influencing the composition of the intestinal microbiota in early infancy. *J. Pediatr.* 118(2), 511-21.

Queenan, A. M., Bush, K., 2007. Carbapenemases: the Versatile β-Lactamases. *Clin. Microbiol. Rev.* 20(3), 440-458.

Rolain, J-M., 2013. Food and human gut as reservoirs of transferable antibiotic resistance encoding genes. *Front. Microbiol.* 4:173, PMC3690338.

van Schaik, W., 2015. The human gut resistome. *Phil. Trans. R. Soc. B.* 370(1670), 20140087.

WHO, 2017. http://www.who.int/mediacentre/news/releases/2017/bacteria-antibiotics-needed/en/

Zbořil, V., Prokopová, L., Hertlová, M. *Microbiota of gastrointestinal tract – clinical relations* (in Czech) *Mikroflóra tráviaceho traktu – klinické súvislosti.* ISBN 80-247-0584-2, Praha: Grada Publishing, Czech Republic, 2005.

Biographical Sketches

Lucia Birošová

Affiliation: Department of Nutrition and Food Quality Assessment, Faculty of Chemical and Food Technology, Slovak University of Technology

Education: In 2004 she graduated in Biochemistry and Biotechnology at the Faculty of Chemical and Food Technology STU in Bratislava, Slovakia and gained Ing. (MSc). In 2008 she finished her PhD in Microbiology at the Faculty of Chemical and Food Technology STU in Bratislava and gained scientific - academic degree PhD.

Professional Appointments: She works in the area of microbiology for more than 10 years. After PhD studies she has worked as a researcher at the Department of Biochemistry and Microbiology STU. For short time she has also worked at the Department of Experimental and Applied Genetics SMU. Currently, she is an associate professor and deputy head of the Department of Nutrition and Food Quality Assessment. In 2015 she habilitated in Microbiology at the University of Veterinary Medicine and Pharmacy in Košice, Slovakia. Her professional membership is in National Scientific Panel for Biological Hazards, Emerging Risks Expert Network EFSA, Czechoslovak Society for Microbiology. She is a member of the examination board. In 2011 she has gained ČSSM Award "Prize for the best Czech and Slovak microbiology" for her scientific results. In 2013 she has acquired RAJO award for best lecture at FIEP congress in section "Nutrition, movement activities and healthy active life style". Several times, she has also received a FEMS grant (Young Scientist Meeting Grant). Her research is focused mainly on the issue of antibiotic resistance in food and the environment. It also addresses the monitoring of the natural and synthetic substances biological effects, or the study of intestinal microbiota in relation to nutrition. She is the author or co-author of several scientific articles in peer-reviewed scientific journals and chapters in foreign monographs, co-author of two domestic monographs, as well as university textbooks. She participates in various grant projects as a principal investigator or co-worker.

Honors:

- "STU price for significant scientific contribution for research team (2016)"
- ČSSM Award "Prize for the best Czech and Slovak microbiology"
- RAJO award for best lecture at FIEP congress in section "Nutrition, movement activities and healthy active life style"
- 5x FEMS grant (Young Scientist Meeting Grant)

Publications from the Last 3 Years:

1. Lépesová, K., Mackuľak, T., Birošová, L.: Effect of wastewater pollution on the development and spread of antibiotic resistance in bacteria, *Chemické listy* 111(6), 374-380, 2017 (IF 0,387)
2. Osziová, H., Bírošová, L.: Study of the sensitivity of micro-organisms to biocides, *Chemické listy* 111(3), 217-222, 2017 (IF 0,387)
3. Olejníková, P., Kaszonyi, A., Šimkovič, M., Lakatoš, B., Kaliňák, M., Valachovičová, M., Birošová, L.: Differences in gut micro-biota activity (antimicrobials, potential mutagens, and sterols) according to diet, *Acta Alimentaria* 46(1), 61-68, 2017

4. Brandeburova, P., Bírošová, L., Vojs, M., Kromka, A., Gál, M., Tichý, J., Híveš, J., Mackuľak, T.: The influence of selected nanomaterials on microorganisms, *Monatsh Chem.* 148(3), 525-530 2017
5. Mackuľak, T., Marton, M., Radičová, M., Vojs Staňová, A., Grabic, R., Bírošová, L., Nagyová, K., Vojs, M., Bodík, I., Brandeburová, P., Gál, M.: Monitoring of micropollutants and resistant bacteria in wastewater and their effective removal by boron doped diamond electrode, *Monatsh Chem.* 148(3), 539-548, 2017
6. Zelená, V., Birošová L., Olejníková, P., Polák, M., Lakatoš, B., Varečka, Ľ.: Properties of anaerobic fungi isolated from several habitats: complexity of phenotypes, *General Physiology and Biophysics* 35(1), pp. 95-107, 2016
7. Šaková, N., Olejníková, P., Kaliňáková, B., Pangallo, D., Lakatoš, B., Nagyová, K., Valachovičová, B., Birošová, L.: Comparative study of adult Slovak vegetarians and meat-eaters gut microflora, *J Food Nutrition Res* 54(3), pp. 260-269, 2015
8. Olejníková, P., Birošová, L., Švorc, L., Vihonská, Z., Fiedlerová, M., Marchalín, Š., Šafář, P.: Newly synthesized indolizine derivatives - antimicrobial and antimutagenic properties, *Chem. Pap.* 69(7), 983-992, 2015
9. Mackuľak, T., Birošová, L., Gál, M., Bodík, I., Grabic, R., Ryba, J., Škubák, J.: Wastewater analysis: the mean of the monitoring of frequently prescribed pharmaceuticals in Slovakia, *Environmental Monitoring and Assessment* 188 (1), 1 -12, 2016
10. Mackuľak, T., Vois, M., Grabic, R., Golovko, O., Vojs Staňová, A., Birošová, L., Medveďová, B., Híveš, J., Gál, M., Kromka, A., Hanusová, A.: Occurrence of pharmaceuticals, illicit drugs, and resistant types of bacteria in hospital effluent and their effective degradation by boron-doped diamond electrodes, *Monatsh Chem.* 147(1), pp. 97-103, 2016
11. Mackuľak, T., Birošová, L., Bodík, I., Grabic, R., Takáčová, A., Smolinská, M., Hanusová, A., Híveš, J., Gál, M.: Zerovalent iron

and iron (VI): effective means for the removal of psychoactive pharmaceuticals and illicit drugs from wastewaters, *Sci. Tot. Env.* 539 art. No. 18328, pp. 420-426, 2016

12. Mackuľak, T., Grabic, R., Gál, M., Gál, M., Birošová, L., Bodík, I.: Evaluation of different smoking habits during music festivals through wastewater analysis, *Env. Tox. Pharm.*40 (3), pp. 1015-1020, 2015
13. Mackuľak, T., Nagyova K., Grabic, R., Koba, O., Gál, M., Birošová, L.: Utilization of Fenton-like reaction for antibiotics and resistant bacteria elimination in different parts of WWTP, *Env. Tox. Pharm.* 40(2), pp. 492-497, 2015
14. Mackuľak, T., Birošová, L., Grabic, R., Škubák, J., Bodík, I: National monitoring of nicotine use in Czech and Slovak Republic based on wastewater analysis, *Env. Sci. Pol. Res.* 22(18), pp. 14000-14006, 2015
15. Mackuľak, T., Mosný, M., Škubák, J., Grabic, R., Birošová, L.: Fate of psychoactive compounds in wastewater treatment plant and the possibility of their degradation using aquatic plants, *Env. Tox. Pharm.* 39(2), 969-973, 2015
16. Mackuľak, T., Mosný, M., Grabic, R., Golovko, O., Koba, O., Birošová, L.: Fenton-like reaction: A possible way to efficiently remove illicit drugs and pharmaceuticals from wastewater, *Env. Tox. Pharm.* 39 (2), 483-488, 2015
17. Mackuľak, T., Škubák, J., Grabic, R., Ryba, J., Birošová, L., Fedorova, G., Špalková, V., Bodík, I.: National study of illicit drug use in Slovakia based on wastewater analysis, *Sci. Tot. Env.* 494-495, 158-165, 2014
18. Birošová L., Mackuľak, T., Bodík, I., Ryba, J., Škubák, J., Grabic, R.: Pilot study of seasonal occurrence and distribution of antibiotics and drug resistant bacteria in wastewater treatment plants in Slovakia, *Science of the Total Environment* 490, 440-444, 2014
19. Žemlička, L., Fodran, P., Lukeš, V., Vagánek, A., Slováková, M., Staško, A., Dubaj, T., Karabín, M., Birošová, L., Rapta, P.:

Physicochemical and biological properties of luteolin-7-O-glucoside (cynaroside) isolated from Anthriscus sylvestris (L.) Hoffm, *Monatsh Chem* 145 (8), 1307-1318, 2014

Karolína Kislíková

Research and Professional Experience: She has completed the first level of higher education by studying the accredited Bachelor´s study programme Nutrition, Cosmetics, Health Protection in the study field Food Science and Technology at the Faculty of Chemical and Food Technology STU in Bratislava and she was granted the Academic Title of abbreviated as BC.

In the present, she continues in the second level of higher education at the Chemistry and Technology of Foodstuff, Branch: Food Science and Biotechnology at the Faculty of Chemistry, Brno University of Technology, Czech Republic.

Research: Her research was to study intestinal microbiota, monitoring and establish a ratio of antibiotic resistant coliform bacteria and enterococci in stools of healthy people. She is the author of bachelor thesis: Monitoring of antibiotic resistant bacteria prevalence in human gut.

Kristina Lépesová

Affiliation: Department of Nutrition and Food Quality Assessment, Faculty of Chemical and Food Technology, Slovak University of Technology, Bratislava, Slovakia

Education: In 2014 she graduated in Nutrition and Health Protection at the Faculty of Chemical and Food Technology STU in Bratislava, Slovakia and gained Ing. (MSc)

Professional Appointments: She works in the area of microbiology for more than 5 years. After she has completed the second level of higher education with honors, she has log on the PhD study in the field of Biochemistry. In 2015 she has received a FEMS grant (Young Scientist Meeting Grant) as well as prestigious award in the form of conference passport by BASF Ltd. at interactive conference PREVEDA. Her research is focused mainly on the issue of antibiotic resistance in food and the environment, monitoring of the natural and synthetic substances biological effects, or the study of intestinal microbiota in relation to nutrition. She is the author or co-author of several scientific articles in peer-reviewed scientific journals and she participates in various grant projects as a principal investigator or co-worker. She has also presented her results several times at domestic or international conferences.

Honors:

- FEMS grant (Young Scientist Meeting Grant)
- Conference passport by BASF Ltd.
- 2x principal investigator of STU grant for young scientists

Publications from the Last 3 Years:

1. Lépesová, K., Mackuľak, T., Bírošová, L., 2017. Effect of Wastewater Pollution on the Development and Spread of Antibiotic Resistance in Bacteria. *Chemické listy* 111 (6), 374-380.
2. Mackuľak, T., Marton, M., Radičová, M., Vojs Staňová, A., Grabic, R., Birošová, L., Nagyová, K., Vojs, M., Bodík, I., Brandeburová, P., Gál, M., 2016. Monitoring of micropollutants and resistant bacteria in wastewater and their effective removal by boron doped diamond electrode. *Monatschefte fur Chemie*, DOI 10.1007/s00706-016-1914-4.
3. Šaková, N., Olejníková, P., Kaliňáková, B., Pangallo, D., Lakatoš, B., Nagyová, K., Valachovičová, M., Birošová, L., 2015. Comparative study of adult Slovak vegetarians and meat-eaters gut

microbiota. *Journal of Food and Nutrition Research* 54, ISSN 1336 – 8672, 260 – 269.

4. Mackuľak, T., Nagyová, K., Fáberová, M., Grabic, R., Koba, O., Gál, M., Birošová, L., 2015. Utilization of Fenton–like reaction for antibiotics and resistant bacteria elimination in different parts of WWTP. *Environmental Toxicology and Pharmacology* 40, 492 – 497.
5. Hybenová, E., Birošová, L., Nagyová, K., Štofirová, J., Šaková, N., Olejníková, P., Kaliňáková, B., 2014. Testing of selected probiotic properties of lactic acid bacteria isolated from vegetarians and meat-eaters faeces. *Acta Chimica Slovaca* 7, 38— 43.

In: Nutrients, Wastewater and Leachate
Editor: Amimul Ahsan
ISBN: 978-1-53613-949-5

Chapter 5

LANDFILL LEACHATE TREATMENT BY ELECTROLYSIS AND PREPARATION OF LOW COST ACTIVATED CARBON

Iqbal K. Erabee[1,2], Amimul Ahsan[3,4,*], Monzur A. Imteaz[4] and A. W. M. Ng[5]

[1]Department of Civil Engineering, Faculty of Engineering, University Putra Malaysia, Selangor, Malaysia

[2]Department of Civil Engineering, Faculty of Engineering, University of Thi-Qar, Iraq

[3]Department of Civil Engineering, Uttara University, Dhaka, Bangladesh

[4]Department of Civil and Construction Engineering, Swinburne University of Technology, Melbourne, Australia

[5]College of Engineering and Science, Victoria University, Melbourne, Australia

[*] Corresponding Author Email: ashikcivil@yahoo.com.

Abstract

This study investigated the effects of different parameters on the removal efficiencies of organic and inorganic pollutants in landfill leachate treatment by electrolysis. Different parameters were considered such as the electric potential (e.g., 24, 40 and 60 V), hydraulic retention time (HRT) (e.g., 40, 60, 80, 100 and 120 min), sodium chloride (NaCl) concentration (e.g., 1, 3, 5 and 7%), pH (e.g., 3, 7 and 9), electrodes materials (e.g., aluminum (Al) and iron (Fe)) and distance between electrodes (e.g., 1, 2 and 3 cm).

The best operational condition of electrolysis was then recommended. The electric potential of 60 V with HRT of 120 min at 5% of NaCl solution using Al as anode and Fe as cathode (kept at a distance of 3 cm) was the most efficient condition which increased the removal efficiencies of various parameters such as turbidity, salinity, total suspended solids (TSS), total dissolved solids (TDS), biochemical oxygen demand (BOD), chemical oxygen demand (COD) and heavy metals (e.g., Zn and Mn). The higher removal percentages of many parameters, especially COD (94%) and Mn (93%) indicated that the electrolysis is an efficient technique for multi-pollutants (e.g., organic, inorganic and heavy metals) removal from the landfill leachate.

Two raw materials, sago palm bark (SPB) and date pits, were utilized as precursors to prepare high porosity activated carbon (AC). The porosity of these two raw materials was compared with that of commercial AC made from coconut shells. The physicochemical activation method was used for AC preparation, and it consisted of two steps, carbonization and activation. The activation process was performed using zinc chloride ($ZnCl_2$) as an activation agent. N_2 adsorption-desorption analysis was carried out to characterize the porosity of AC. Thermogravimetric analysis (TGA) was conducted for the two raw materials. The adsorbent made from SPB, which showed the maximum surface area of 1634 m^2/g at the 700°C activation temperature for one hour, while the surface area of prepared AC from date pits was 1367 m^2/g. Both prepared ACs had a larger surface area than commercial AC made with coconut shell (1348 m^2/g).

Nomenclature

AC	Activated carbon
BOD	Biochemical Oxygen Demand (mg/L)

COD	Chemical Oxygen Demand (mg/L)
HRT	Hydraulic retention time (min)
SPB	Sago palm bark
TDS	Total dissolved solid (mg/L)
TGA	Thermogravimetric analysis
TSS	Total suspended solid (mg/L)

1. INTRODUCTION

Leachate is a type of wastewater, which is generated when water infiltrates through or is exposed to the waste accumulation. Leachate has high concentrations of organic as well as inorganic components especially in the immediate vicinity of the landfill. It might comprise suspended material such as waste materials, chemical precipitates, and bacteria colonies, as well as dissolved chemicals including sodium, chloride, aluminum, and iron. However, there are multiple concerns that make the process of selecting a proper treatment option quite challenging and extremely site-specific. To eliminate pollutants from leachate, an extensive range of treatment and filtration alternatives are available. The selection of the suitable approach is contingent on many technical aspects. These include the volume and quality of leachate, design of landfill, extent of treatment required, and final disposal means for discharges and residues, as well as non-technical aspects like legal matters. Passive treatment mechanisms were devised based on several similar abovementioned treatment systems.

Treatment of landfill leachate using electrolysis process is a modern technology in which a sacrificial metal anode and cathode produces electrically active coagulants and tiny bubbles of hydrogen and oxygen in water. Electrolysis process has the advantage of its easiness in setup and operation. Some challenges are being faced during the process like formation of sludge which acts as an insulating deposit on the electrode and lower efficiency [1].

Electrochemical oxidation of pollutants in wastewater is potentially accomplished by two distinct mechanisms, e.g., indirect oxidation and direct anodic oxidation. In indirect oxidation, a mediator is electrochemically generated to accomplish the oxidation whereas, in direct anodic oxidation, the pollutants are destroyed at the anode surface. In indirect oxidation, agents are generated anodically and are responsible for oxidation of inorganic and organic pollutants [2]. The electrolyzed wastewater or leachate can be polished further using activated carbon. A waste material converted to activated carbon is a great potential form of waste recycling [3-4]. A few investigators studied on the characterization of activated carbon [5-7] and on utilization of it in wastewater and leachate treatment [8-9].

The electrolyte, direct current supply and the pair of electrodes are generally associated with the electrolysis process [10]. Cathode material is usually stainless steel, except copper which was used by Tsai et al. [11] and titanium used by Moraes and Bertazzoli [12]. While Othman et al. [13] used two monopolar Fe plate as electrodes (anodes and cathodes). Tsai et al. [11] employed Al and Fe (anodes) to simultaneously utilize electro-coagulation, which is responsible for the removal of organics with high molecular weight. Other important factor that affects electrolysis oxidation (EO) is the gap between electrodes. Chiang et al. [14] set the electrode gap at 1.5 cm during batch electro-oxidation mode and 0.5 cm during continuous electro-oxidation mode. Song et al. [15] found that 96% of color and 80% total organic carbon (TOC) can be removed by keeping the electrodes at a distance of 3 cm.

Recently, many studies included the effect of electric potential and pH on the removal efficiencies of organic and inorganic contaminants in wastewater. Belkacem et al. [16] reported that more than 90% BOD, COD, total suspended solids (TSS) and color were removed by using Al electrodes at 1 cm gap, 20 V electric potential and 20 min of hydraulic retention time (HRT). Mahvi et al. [17] concluded that the best sulfate removal was obtained at 30 V of electric potential, 60 min of HRT and pH of 11. Bukhari [18] investigated the removal of TSS and turbidity from municipal wastewater using electro-coagulation treatment process with stainless steel electrode. The removal efficiency of TSS was 95.4% with a current of 0.8A and 5 min of HRT. Nouri et al. [19] concluded that the best removal of chromium (Cr (VI)) ions was investigated by using Fe electrodes at pH of 3, 20-60 min of HRT and electrical potential of 40 V. Shafaei et al. [20] concluded that Mn^{+2} ions could be removed in high percentage by electrocoagulation with Al electrodes under optimum pH of 7.

Curteanu et al. [21] used indirect electrolysis for the removal of chlorophyll, TSS, and COD from the final effluent of aerated lagoons in wastewater treatment plant. The efficiency of the process was studied experimentally and by simulation using neural networks at different conditions (retention time of 5 to 50 min using two types of electrodes (Al and stainless steel) at different distances (from 1 to 3.5 cm) with electric potential of 50 V). By using stainless steel electrodes with 30 min retention time, the removal efficiencies of TSS, COD and chlorophyll were around 38, 41 and 64%, respectively. By applying Al electrodes, the removal efficiencies were 97, 94 and 0.5% for TSS, COD and chlorophyll, respectively. Tao et al. [22] tested the removal of heavy metals from fly ash leachate using a novel combined approach by bio-electrochemical systems followed by electrolysis reactors. The metal removal efficiencies were 98.5, 95.4 and 98.1% for copper, zinc and lead, respectively. Moro et al. [23] employed a batch-type electrolytic cell to treat raw leachate for 240 min at a current density of $200mA/cm^2$. The electrochemical oxidation found to be effective in removal of ammonia nitrogen (NH_3-N) by 82% at the end of the test.

The adsorption process is mainly used in the treatment of wastewater as a secondary treatment using adsorbent with high surface area such as activated carbon. Activated carbon (AC) is prepared to have very large surface area so that it can be widely applied in applications such as the treatment of water and wastewater and sanitary and hazardous landfill leachates. AC is produced from carbonaceous materials such as palm oil wastes [24], coconut shell [25], sugarcane bagasse [26], rice husks [27], and raw wheat residues [28]. Many researchers studied on activated carbon characterization [29, 30, 31] and on utilization of it [32, 33].

Different methods used to prepare AC involve physical, chemical, and physicochemical activation processes. Physical methods involve carbonization and activation at elevated temperature using an activating agent such as steam, air, CO_2, or their mixtures. Physicochemical activation is a new method to prepare AC that combines physical and chemical methods [34].

The sago palm belongs to the Palmae family and the genus *Metroxylon*. The sago palm grows well in humid tropical climates such as those of Oceania and Southeast Asia. It can reach a maximum height of 25 m and a diameter of 40 cm. In Malaysia, the area with the largest sago plantations is located outside the Peninsula and is found in the state of Sarawak [35]. The sago starch industry discards over 20,000 tons of SPB per year. SPB is Malaysia's main carbohydrate. It is about 60 to 70% cellulose and hemicellulose [36].

The purpose of this study is to investigate the influence of electrical potential (e.g., 24, 40 and 60V), HRT (e.g., 40, 60, 80, 100 and 120 min), NaCl concentration, pH (e.g., 3, 7 and 9), electrodes materials (e.g., Al and Fe) and distance between electrodes (e.g., 1, 2 and 3 cm) on pollutants removal efficiency in landfill leachate treatment by electrolysis. Finally, the best operational condition of electrolysis is recommended.

The second objective of the study is to examine the porosity of AC prepared from two different raw materials; sago palm bark and date pits. These materials were compared with commercial AC made from coconut shells.

2. Methodology and Materials of Electrolysis Treatment

The samples of landfill leachate were collected from the Jeram Sanitary Landfill (JSL) at a pre-determined discharge point where the samples represent the liquid generated from three main pipelines in the area currently in operation. The sample was collected in plastic containers, transferred to the laboratory, and stored at 4°C. Laboratory analysis of leachate sample characteristics was carried out based on standard methods for the examination of water and wastewater [37]. The effluent discharge standards (Standards B) stated in the Sewage and Industrial Effluents Regulations 1979 by Department of Environment, Malaysia are shown in Table 1. The measurements of pH, total dissolved solids (TDS), salinity and electrical conductivity were performed using a Tracer Pocket tester (Code: 1766, LaMotte, Taiwan). The TSS was determined by pouring a well-mixed sample (20 mL) through a pre-weighted filter paper that was mounted on a vacuum filtration apparatus. Standard method [37] was used to measure the COD and BOD of the sample. A multidetector optical turbid meter was used to test the turbidity.

Table 1. Initial characteristics of raw leachate taken from JSL [38]

Parameters	Unit	Value	Malaysian Standards*
pH	-	8.45	5.5 – 9.0
TDS	ppm	15566	-
TSS	mg/L	970	100
Salinity	ppm	23300	-
Turbidity	NTU	768	-
E. Conductivity	μS/cm	41.80	-
COD	mg/L	6400	100
BOD	mg/L	690	50
Zn	mg/L	2.70	1.0
Mn	mg/L	5.00	1.0

*Sewage and Industrial Effluents (Standards B), 1979, DOE Malaysia.

Heavy metals like zinc (Zn) and manganese (Mn) were measured using a multi-detector optical colorimeter (HACH DR/890 Colorimeter, Loveland, CO, USA). Electrolysis treatment process A batch electrolytic reactor (Figure 1) was made up of glass container with dimensions of 300 mm length, 200 mm width and 250 mm height, and two pairs of electrodes placed (at a distance of 1, 2 and 3 cm apart). In the first step, Al was used for cathode and anode with dimensions of 39.5 cm length, 2.5 cm width and 0.1 cm thickness. While in the second step, Al was used as anode and

Fe as cathode with dimensions of 50 cm length, 3.5 cm width and 0.5 cm thickness. The electrode sets were arranged parallel to each other and connected to the negative and positive outlets of the digital Direct Current (DC) power supply (Model RXN-3020 D) that regulates the electricity. The pairs of electrodes in each step were submerged in the electrolysis reactor which contained leachate sample of 3 L. An aerator with an average flow of 1200 cc/min was placed on the bottom of the reactor in order to achieve proper mixing in the leachate samples. The electrodes were subjected to electric potentials of 24, 40 and 60 V from the DC power converter. In each value of electric potential, the sample of electrolyzed leachate was taken in different HRTs of 40, 60, 90 and 120 min. The initial values of electric potential and HRT were 24 V and 40 min, respectively. Two trials were performed in order to investigate the performance of electrode materials on higher RE of organic and inorganic pollutants. At first, electrolysis was conducted using Al in both anode and cathode and in the second trial; Al was used as anode and Fe as cathode. The parameters tested were electrical conductivity, salinity, turbidity, COD, BOD, TSS, TDS and heavy metals (Zn and Mn). Also, the effect of different distances between electrodes (1, 2 and 3 cm) was investigated in the present study.

2.1. Preparation of Electrolyzed Sample with Different Ph

Electrolyzed leachate samples were prepared at the electrolysis cell with different initial pH values of 3, 7 and 9 were prepared by adding an appropriate amount of 0.1 N of HCl prior to start each experiment.

The removal efficiency (RE) of various parameters is calculated using Equation 1.

$$RE\ (\%) = \frac{(C_0 - C_e)}{C_0} \times 100 \qquad (1)$$

where, C_0 = initial concentration (mg/L) and C_e = final concentration (mg/L).

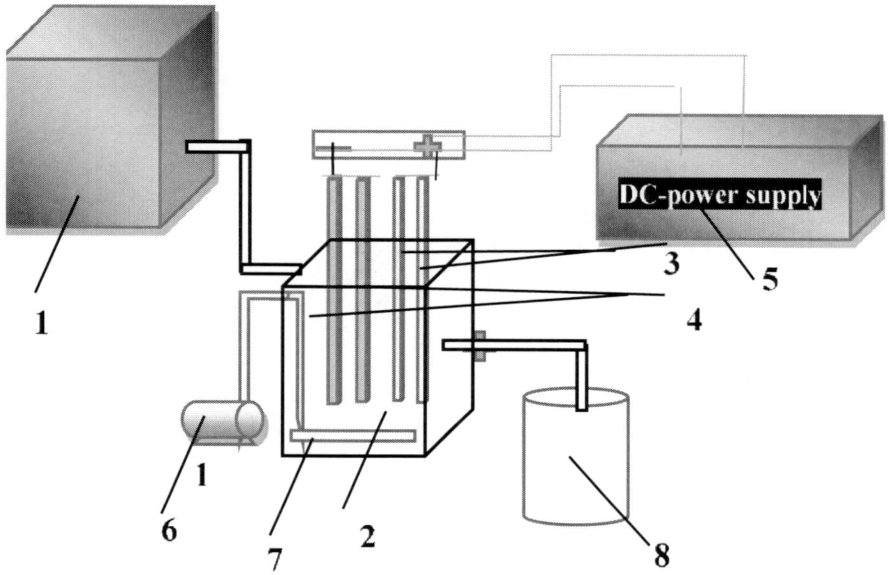

Figure 1. Schematic of electrolysis cell: (1) raw leachate storage tank (5 L); (2) electrolysis reactor; (3) pair of Anode; (4) pair of cathode; (5) DC power supply; (6) aeration motor; (7) perforated tube; and (8) effluent storage tank [38].

2.2. Effect of Electric Potential and HRT

The electrolysis treatment of leachate sample was conducted at different voltages (24, 40 and 60 V) and HRTs of 40, 60, 90 and 120 min. The results obtained after the electrolysis indicated that there was decrease in the final concentration (Ce) (of parameters) with increase of HRT for the three different voltages applied to the electrodes. Figure 2 shows the removal percentage of various parameters with different voltages and HRTs. The maximum RE for TDS, TSS, COD, BOD, salinity, turbidity, Zn and Mn were 78, 82, 94, 87, 82, 87, 87 and 93% respectively at HRT of 120 min. It can be concluded that for the given range of electric potential (24 to 60 V) and HRT (40 to 120 min), the removal efficiency increased with increasing of HRT and electric potential.

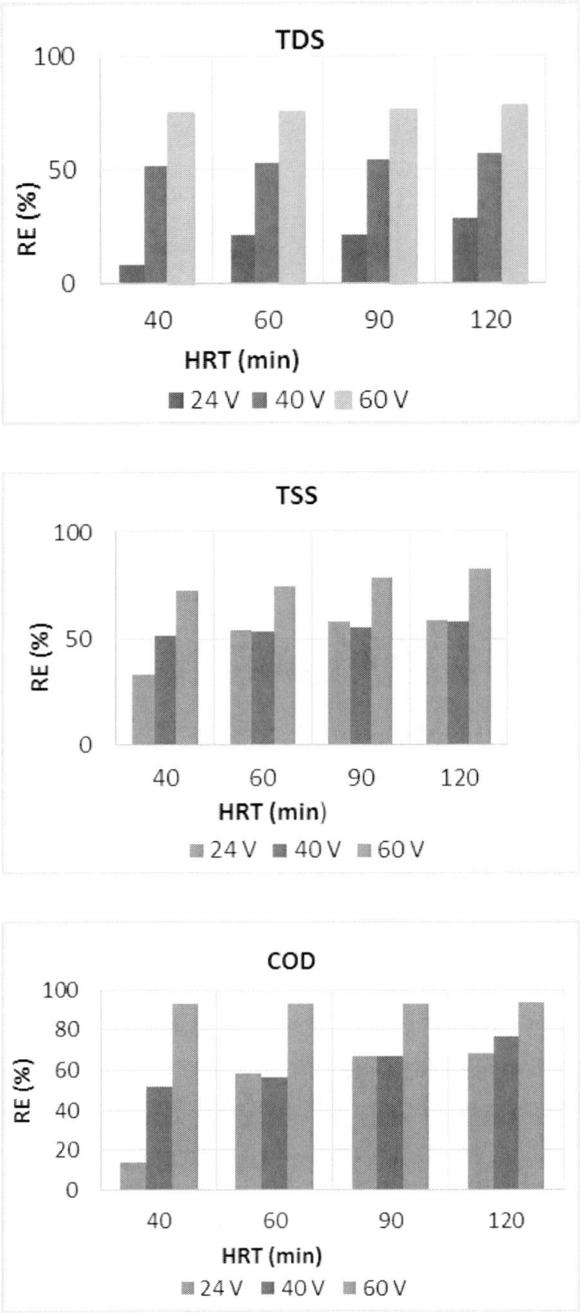

Figure 2. (Continued)

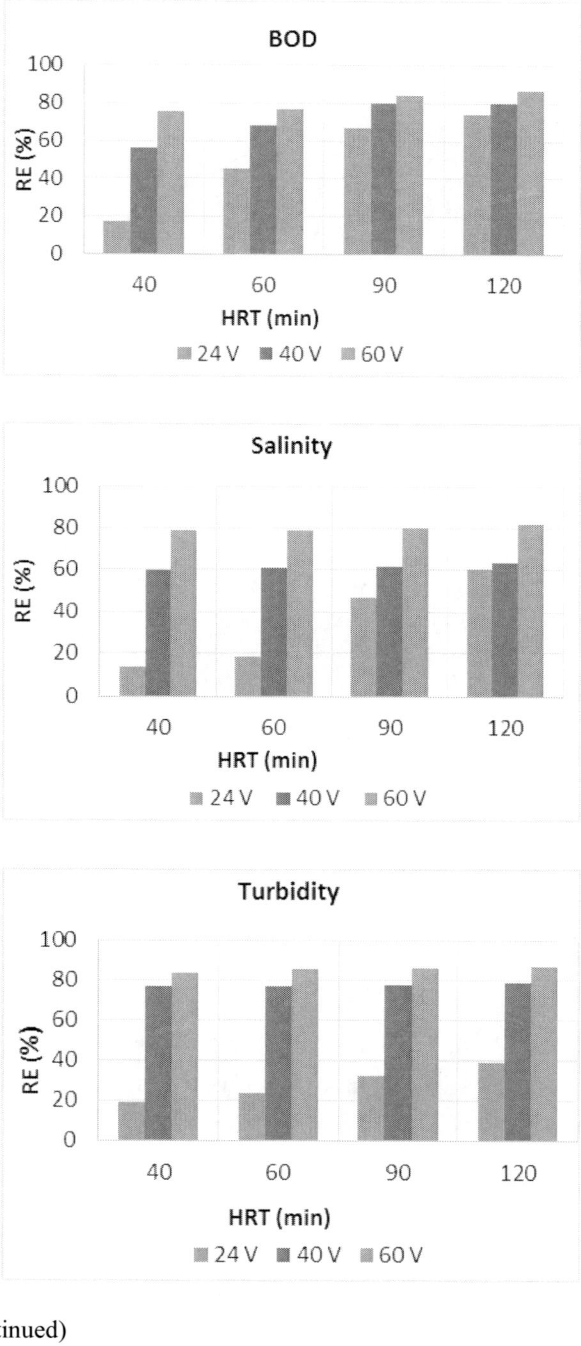

Figure 2. (Continued)

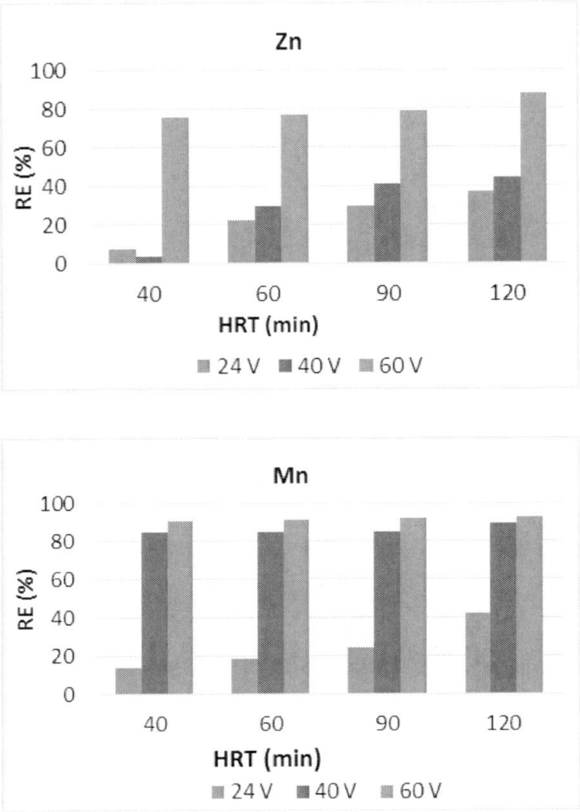

Figure 2. Removal efficiency at different voltages and HRTs [38].

2.3. Effect of pH

Different pH values (3, 7 and 9) were obtained by adding drops of 0.1 N HCl to the samples of leachate. Ce was calculated for each parameter after 2 h of treatment at 60 V. RE of various parameters at different pH are shown in Table 2, respectively. The lowest Ce and the highest RE were obtained when pH was 9 for all the parameters except Zn. In the case of Zn, the lowest Ce and highest RE were obtained when pH was 3. The highest Ce and the lowest RE were obtained when pH was 7 except for COD. Which means that the maximum removal efficiency of each parameter was obtained in electrolyzed solution with pH=9.

Table 2. RE (%) of various parameters at different initial values of pH [38]

pH	Salinity	Turbidity	TDS	TSS	COD	BOD	Zn	Mn
3	23.18	5.46	7.49	7.21	88.93	67.06	85.19	55.60
7	3.86	36.72	3.64	2.06	89.56	23.14	66.67	22.60
9	82.19	85.68	77.52	82.47	93.75	85.47	79.63	70.00

2.4. Effect of Electrode Material

The study included using two types of materials for electrode, firstly, using Al in both cathode and anode and secondly, using Al as anode and Fe as cathode. Figure 3 shows the removal percentage of various parameters using these two types of electrodes. The maximum RE for the parameters was obtained using Al in anode and Fe in cathode compared to Al at both anode and cathode except for TDS. For TDS both electrode materials give almost the same RE. These results are in agreement with other studies conducted by Emamjomeh and Sivakumar [39] and Ramprasad [40] who achieved the higher removal efficiencies of many parameters like TDS, COD and ammonia by using Al and Fe electrodes in treatment of wastewater.

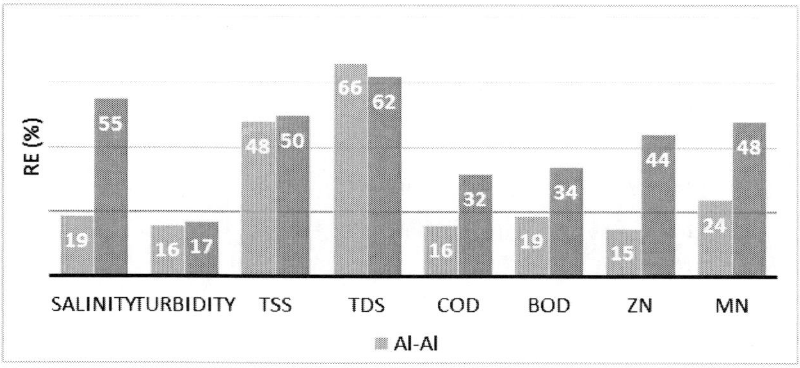

Figure 3. Effect of electrode materials on removal efficiency of various parameters [38].

3. Activated Carbon Adsorption Treatment

3.1. Preparation of AC

Sago palm bark (SPB) and date pits (DP) were used as precursor materials (Figure 4). The SPB was purchased from a local plantation in Melaka, Malaysia. The preparation of the precursor included the removing of pith (the inner portion) from sago palm trunks to obtain the bark fraction (the outer layer), and the collected bark was dried at 105°C for 24 h. After drying, the bark of sago palm was chopped and screened to 20 to 30 mm and stored in sealed plastic bags at 20°C. The small particles were crushed and grinded. Ground samples were sieved to the particle size of 0.3 to 0.6 mm. The DP were washed with tap water several times, dried overnight at 105°C, crushed, and sieved into sizes from 0.3 to 0.6 mm.

The preparation of AC included two steps; carbonization and chemical activation. In carbonization, the sample was transferred to a cylindrical stainless steel reactor. The reactor was inserted into an electric horizontal tubular furnace under continuous nitrogen gas flow (100 mL/min). The furnace tube dimensions are 800 mm length and 50 mm in diameter. The heating started by adjusted the furnace temperature to the desired value. The furnace took about 2 to 2.5 h to reach the desired temperature (the average heating rate was 10°C/min).

The temperature of carbonization was 300°C and the total duration of the reaction was 4 h. The second step of the preparation was the activation using the chemical agent ($ZnCl_2$) with impregnation ratio of 1/1 (weight of $ZnCl_2$ per weight of precursor). The precursor and chemical agent were mixed thoroughly for 24 h, and the sample was then dried overnight at 100°C. After drying, the impregnated material was inserted into a tubular regulated furnace at an activation temperature of 700°C for 1 h under continuous nitrogen flow (100 mL/min). After cooling, the sample was washed with boiling and cold distilled water thoroughly until the pH reached a neutral value. Finally, the sample was dried at 100°C overnight and stored in a desiccator.

(a) SPB (b) DP

Figure 4. Raw materials of AC [3].

3.2. Activated Carbon Characterization

3.2.1. Moisture and Ash Contents

The moisture and ash contents were calculated using the oven-drying test according to ASTM D2867-09 [40] and ASTM D2866-94 [42], respectively.

3.2.2. Yield

$$Yield\,(\%) = \frac{weight\ of\ AC\ after\ carbonisation}{weight\ of\ the\ raw\ material} \times 100 \qquad (2)$$

3.2.3. Ultimate Analysis

The ultimate analysis for measuring the percentage of four elements of raw materials (carbon (C), hydrogen (H), nitrogen (N), and sulfur (S)) was conducted using CHNS analyzer (Elementar Vario El CUBE). While the percentage of oxygen (O) was calculated by the following formula,

$$Oxygen\ (\%) = 100 - (C\% + H\% + N\% + S\%) \tag{3}$$

3.2.4. Thermogravimetric Analysis (TGA)

The TGA was performed using a thermogravimetric analyzer (SDTA851e, Mettler Toledo, Switzerland). The sample was heated from 25 to 800°C at the rate of 10°C/min.

3.2.5. Porosity Characterization

Porosity characterization involves the determination of the surface area of the prepared activated carbon, total pore volume, and pore size distribution through N_2 adsorption-desorption isotherms at -195.6°C using an automatic adsorption instrument. The surface area (S_T) of the sample was evaluated using the BET (Brunauer, Emmett and Teller) method over a relative pressure range of 0.01 to 0.3 [43]. Table 3 presents the proximate and ultimate analysis of these three raw materials. The proximate analysis of SPB and date pits showed low ash content (2.09 and 1.35%, respectively), which is an important factor in the AC preparation method.

Table 3. Ultimate and Proximate Properties of Precursors [3]

Proximate Analysis (%)	[44]		
Property	Coconut Shells	Sago Palm Bark	Date Pits
Moisture content	10.46	9	10.3
Yield	-	65	58
Ash content	3.58	2.09	1.35
Ultimate Analysis (%)	[45]		
Element	Coconut Shells	Sago Palm Bark	Date Pits
C	40.12	45.16	45.63
H	2.56	6.306	7.118
N	0.61	-0.0144	0.724
S	0.23	0.013	0.062
O	56.48	48.54	46.47

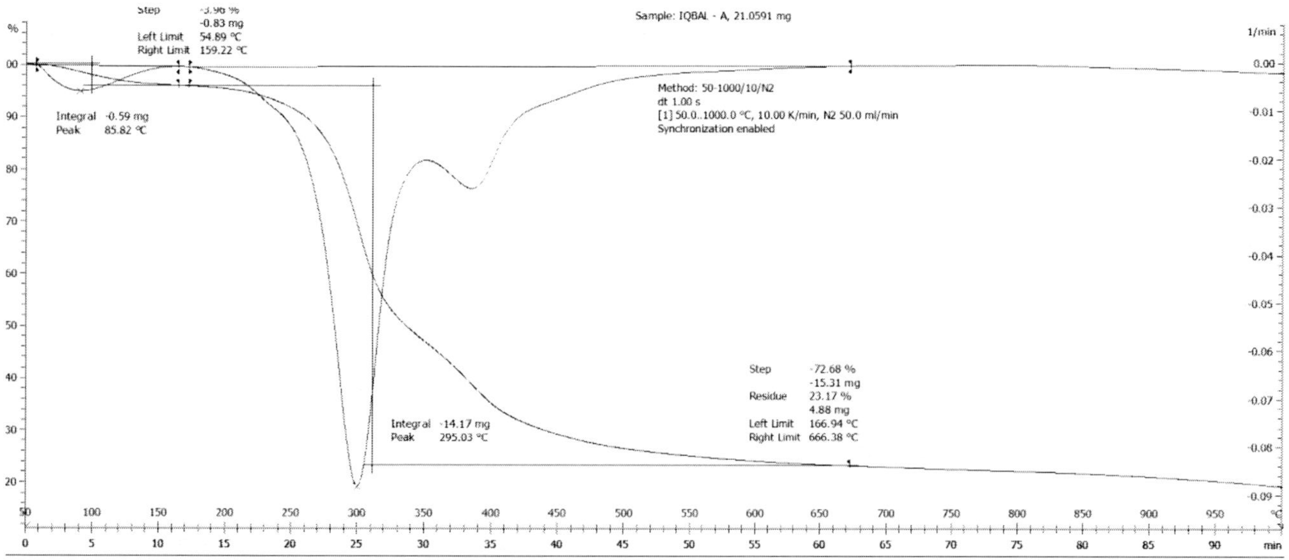

Figure 5. TGA profile of date pits (as an example) [3].

The weight loss of the raw materials with respect to temperature was approximated by TGA. As an example, Figure 5 presents the TGA profile of date pits. The first step represents 3.96% of weight loss in temperatures ranging from 55 to 160°C; this loss represented moisture release. The second step represents weight loss of 72.68% in the temperature range from 167 to 667°C. This weight loss was due to the decomposition of cellulose and other main components. Up to 670°C, slow weight loss was registered due to the decomposition of hemicellulose and lignin. Hence, the carbonization temperature is considered in the range of 300 to 700°C.

3.2.6. Internal Surface Area of AC

The raw precursor materials played a critical role in producing AC with high porosity. Figures 6 and 7 show the porosity characteristics of AC using three different precursors: AC made from sago palm bark (AC-SPB), AC made from date pits (AC-DP), and AC made from coconut shell (AC-CN). The surface area for AC-SPB, AC-DP and AC-CN were 1634, 1367 and 1348 m^2/g, respectively (Figure 6). The pore volume of AC-SPB, AC-DP and AC-CN were 0.649, 0.480, and 0.486 cm^3/g, respectively (Figure 7). The maximum porosity characterization of AC-SPB using $ZnCl_2$ with an impregnation ratio of 1.5/1 at 700°C for a holding time of 1 h may reflect that $ZnCl_2$ acts as a dehydrating agent and inhibits the formation of char, which can clog up sample pores [46].

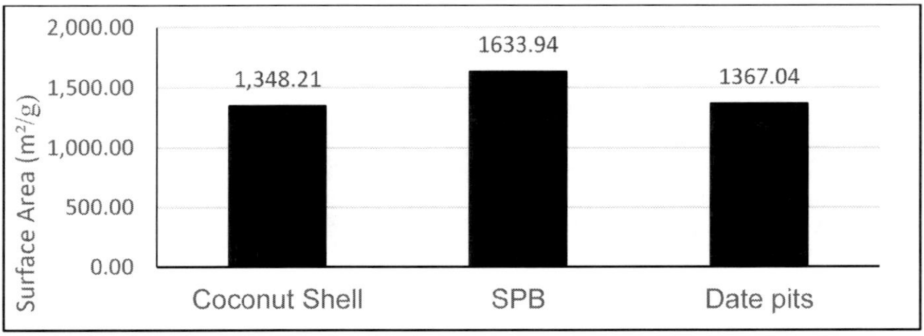

Figure 6. Surface area for the prepared AC using different precursors [3].

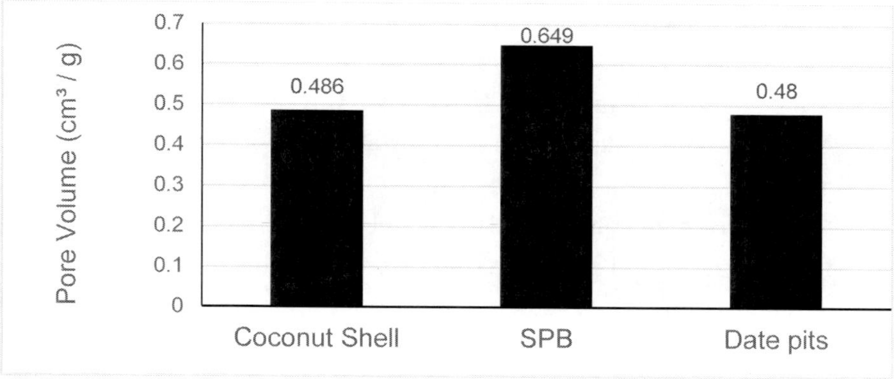

Figure 7. Pore volume for the prepared AC using different precursors [3].

CONCLUSION

In this study, electrochemical oxidation treatment of landfill leachate was applied to increase the removal efficiency of organic and inorganic contaminants, and heavy metals. This leachate is considered as a major source of pollution if it reaches the groundwater table. In the electrolysis reactor, the suspended and colloidal solids get polarized due to the application of electrical field and creating micro aggregates, which then coagulate to form flocks. The electrolysis efficiency was significantly decreased due to the increase in pH and NaCl concentration more than 5%. The best removal efficiency was achieved using Al as anode and Fe as cathode at electrical potential of 60 V with HRT of 120 min. The higher removal percentages of various parameters, especially COD (94%) and Mn (93%) indicated that the electrochemical technology is an efficient solution in landfill leachate treatment. This technique can be applied in treatments of different industrial effluents (e.g., POME), agro-industrial wastewater, latex wastewater, poultry wastewater and municipal wastewater.

Experimental results indicated that adsorbent using date pits with activation by chemical agent of $ZnCl_2$ at 700°C for 1 h resulted in high surface area, which indicates good adsorption for inorganic contaminates and heavy metals when using AC as adsorbent for wastewater treatment.

Development of AC was influenced by different factors such as low ash content of raw material. The two precursors that used in this study have low ash content (2.09 and 1.35%, respectively).

Important factor that affect the preparation method of AC to get good adsorbent is the content of carbon and oxygen. Both precursors prepared in study indicated high content of carbon and oxygen. The activation process of AC with $ZnCl_2$ exhibited maximum porosity characterization such as pore volume, micropore surface area and micropore volume of the prepared AC using SPB in comparison with AC prepared using date pits and commercial AC.

REFERENCES

[1] Chopra, A. K.; Sharma, A. K.; Kumar, V. 2011. Overview of electrolytic treatment: an alternative technology for purification of wastewater. *Archives of Applied Science Research*. Vol *3*(5), 191-206.

[2] Englehardt, J. D.; Deng, Y.; Meeroff, D.; Legrenzi, Y.; Mognol, J.; Polar, J. 2006. *Options for managing municipal landfill leachate: year 1 development of iron-mediated treatment processes*. Florida Center for Solid and Hazardous Waste Management, Florida.

[3] Erabee, I. K.; Ahsan, A.; Daud, N. N. N.; Idrus, S.; Shams, S.; Din, M. F. M.; Rezania, S. 2017. Manufacture of low-cost activated carbon using sago palm bark and date pits by physiochemical activation. *Bio Resources*. Vol *12*(1), 1916-1923.

[4] Ahsan, A.; Kamaludin, M.; Rahman, M. M.; Anwar, A. H. M. F.; Bek, M. A.; Idrus, S. 2014. Removal of various pollutants from leachate using a low cost technique. *Water, Air, and Soil Pollution*. Vol *225*(12), 2163:1-9.

[5] Ahsan, A.; Ismail, N.; Rahman, M. M.; Imteaz, M.; Rahman, A.; Mohammad, N.; Salleh, M. A. M. 2013. Municipal solid waste

recycling in Malaysia: present scenario and future prospects. *Fresenius Environmental Bulletin*. Vol *22*(12a), 3654-3664.

[6] Rahman, M. M.; Bari, Q. H.; Mohammad, N.; Ahsan, A.; Sobuz, H. R.; Uddin, M. A. 2013. Characterization of rice husk carbon produced through simple technology. *Advances in Materials Science and Application*. Vol *2*(1), 25-30.

[7] Shaheed, R.; Azhari, C. H.; Ahsan, A.; Mohtar, W. H. M. W. 2015. Production and characterisation of low-tech activated carbon from coconut shell. *Journal of Hydrology and Environment Research*. Vol *3*(1), 6-14.

[8] Tadda, M. A.; Ahsan, A.; Shitu, A.; El Sergany, M; Arunkumar, T.; Jose, B.; Razzaque, M. A.; Daud N. N. N. 2016. A review on activated carbon: process, application and prospects. *Journal of Advanced Civil Engineering Practice and Research.* Vol *2*(1), 7-13.

[9] Khaleel, M. R.; Ahsan, A.; Imteaz, M.; El-Sergany, M. M.; Daud, N. N. N.; Mohamed, T. A.; Ibrahim, B. A. 2015. Performance of GACC and GACP to treat institutional wastewater: a sustainable technique. *Membrane Water Treatment*. Vol *6*(4), 339-349.

[10] Bae, J. H.; Kim, S. K.; Chang, H. S. 1997. Treatment of landfill leachates: ammonia removal via nitrification and denitrification and further COD reduction via Fenton's treatment followed by activated sludge. *Water Science and Technology*. Vol *36*(12), 341-348.

[11] Tsai, C.; Lin, S.; Shue, Y.; Su, P. 1997. Electrolysis of soluble organic matter in leachate from landfills. *Water Research.* Vol *31*(12), 3073-3081.

[12] Moraes, P. B.; Bertazzoli, R. 2005. Electrodegradation of landfill leachate in a flow electrochemical reactor. *Chemosphere.* Vol *58*(1), 41-46.

[13] Othman, F.; Sohaili, J.; Ni'am, M. F.; Fauzia, Z. 2006. Enhancing suspended solid removal from wastewater using Fe electrodes. *Malaysian Journal of Civil Engineering*. Vol *18*(2), 138-149.

[14] Chiang, H. L.; Huang, C. P.; Chiang, P. C. 2002. The surface characteristics of activated carbon as affected by ozone and alkaline treatment. *Chemosphere*. Vol *47*(3), 257-265.

[15] Song, S.; Yao, J.; He, Z.; Qiu, J.; Chen, J. 2008. Effect of operational parameters on the decolourization of C.I. Reactive Blue 19 in aqueous solution by ozone-enhanced electrocoagulation. *J. Hazard. Mater.* Vol *152*, 204-210.

[16] Belkacem, M; Khodir, M.; Abdelkrim, S. 2008. Treatment characteristics of textile wastewater and removal of heavy metals using the electrolotation technique. *Desalination.* Vol *228*, 245-254.

[17] Mahvi, A. H.; Mansoorian, H. J.; Rajabizadeh, A. 2009. Performance evaluation of electrocoagulation process for removal of sulphate from aqueous environments using plate aluminum electrodes. *World Appl. Sci. J.* Vol *7*, 1526-1533.

[18] Bukhari, A. A. 2008. Investigation of the electro-coagulation treatment process for the removal of total suspended solids and turbidity from municipal wastewater. *Bioresour. Techol*. Vol *99*, 914-921.

[19] Nouri, J; Mahvi, A. H.; Bazrafshan, E. 2010. Application of electrocoagulation process in removal of zinc and copper from aqueous solutions by aluminum electrodes. *Int. J. Environ. Res*. Vol *4*, 201-208.

[20] Shafaei, A.; Rezayee, M.; Arami, M.; Nikazar, M. 2010. Removal of Mn^{+2} ions from synthetic wastewater by electrocoagulation process. *Desalination*. Vol *260*, 23-28.

[21] Curteanu, S.; Piuleac, C. G.; Godini, K.; Azaryan, G. 2011. Modelling of electrolysis process in wastewater treatment using different types of neural networks. *Chemical Engineering Journal*. Vol *172*(1), 267-276.

[22] Tao, H. C.; Lei, T.; Shi, G.; Sun, X. N.; Wei, X. Y.; Zhang, L. J.; Wu, W. M. 2014. Removal of heavy metals from fly ash leachate using combined bioelectrochemical systems and electrolysis. *Journal of Hazardous Materials*. Vol *264*, 1-7.

[23] Del Moro, G.; Prieto-Rodríguez, L.; De Sanctis, M.; Di Iaconi, C.; Malato, S.; Mascolo, G. 2016. Landfill leachate treatment: comparison of standalone electrochemical degradation and combined with a novel biofilter. *Chemical Engineering Journal.* Vol *288*, 87-98.

[24] Ahmad, T., Danish, M., Rafatullah, M., Ghazali, A., Sulaiman, O., Hashim, R., Ibrahim, M. N. M. 2012. The use of date palm as a potential adsorbent for wastewater treatment a review: *Environmental Science and Pollution Research.* Vol 19 (5), 1464-1484.

[25] Bhatnagar, A. M., Choi, Y. H., Jung, W., Lee, S. H., Kim, S. J., Lee, G., Suk, H. S., Kim, B. M., Kim, S. H., Jeon, B. H., and Kang, J. W. 2008. Removal of nitrate from water by adsorption onto zinc chloride treated activated carbon. *Journal of Science and Technology*. Vol 43, 886-907.

[26] Orlando, U. S., Baes, A. U., Nishijima, W., and Okada, M. 2002. A new procedure to produce lignocellulosic anion exchangers from agricultural waste materials, *Bioresources Technology. Vol* 83, 195-198.

[27] El-Shafey. 2007. Sorption of Cd(II) and Se(IV) from aqueous solution using modified rice husk. *Journal of Hazardous Materials. Vol* 147, 319-327.

[28] Wang, Y., Gao, B. Y., Yue, W. W., and Yue, Q. Y. 2007. Adsorption kinetics of nitrate from aqueous solutions onto modified wheat residue. *Colloids and Surfaces A: Physicochemical and Engineering Aspects.* Vol 308(1-3), 1-5.

[29] Rahman, M. M., Bari, Q. H., Mohammad, N., Ahsan, A., Sobuz, H. R., Uddin, M. A. 2013. Characterization of rice husk carbon produced through simple technology. *Advances in Materials Science and Application. Vol* 2(1) 25-30.

[30] Shaheed, R., Azhari, C. H., Ahsan, A., Mohtar, W. H. M. W. 2015. Production and characterisation of low-tech activated carbon from coconut shell. *Journal of Hydrology and Environment Research.* Vol 3(1), 6-14.

[31] Tadda, M. A., Ahsan, A., Shitu, A., ElSergany, M, Arunkumar, T, Jose, B., Razzaque, M. A., Daud N. N. N. 2016. A review on activated carbon: process, application and prospects. *Journal of Advanced Civil Engineering Practice and Research*. Vol 2(1):7-13.

[32] Ahsan, A., Kamaludin, M., Rahman, M. M., Anwar, A. H. M. F., Bek, M. A., Idrus, S. 2014. Removal of various pollutants from leachate using a low cost technique: integration of electrolysis with activated carbon contactor. *Water, Air, and Soil Pollution. Vol* 225(12), 2163:1-9.

[33] Khaleel, M. R., Ahsan, A., Imteaz, M., El-Sergany, M. M., Daud, N. N. N., Mohamed, T. A., Ibrahim, B. A. 2015. Performance of GACC and GACP to treat institutional wastewater: a sustainable technique," *Membrane Water Treatment*. Vol 6(4), 339-349.

[34] Aber, S., Khataee, A., and Sheydaei, M. 2009. Optimization of activated carbon fiber reparation from kenaf using K_2HPO_4 as chemical activator for adsorption of phenolic compounds. *Bioresource Technology*. Vol 100(24), 6586-6591.

[35] Singhal, R. S., Kennedy, J. F., Gopalakrishnan, S. M., Kaczmarek, A., Knill, C. J., and Akmar, P. F. 2008. Industrial production, processing, and utilization of sago palm-derived products. *Carbohydrate Polymers*. Vol 72(1), 1-20.

[36] Ethaib, S., Omar, R., Mazlina, M. K. S., Radiah, A. B. D., and Syafiie, S. 2016. Microwave-assisted dilute acid pretreatment and enzymatic hydrolysis of sago palm bark. *BioResources*. Vol 11(3), 5687-5702.

[37] APHA. *Standard methods for the examination of water and wastewater*. 1998. 20th ed., American Public Health Association, Washington DC, USA.

[38] Erabee, I. K., Ahsan, A., Jose, B., Arunkumar, T., Sathyamurthy, R., Idrus, S., & Daud, N. N. (2017). Effects of electric potential, NaCl, pH and distance between electrodes on efficiency of electrolysis in landfill leachate treatment. *Journal of Environmental Science and Health, Part A,* 1-7.

[39] Emamjomeh, M. M.; Sivakumar, M. 2009. Review of pollutants removed by electrocoagulation and electrocoagulation/flotation processes. *Journal of Environmental Management.* Vol *90*(5), 1663-1679.

[40] Ramprasad C. 2012. Electrochemical treatment of landfill leachate. International *Journal of Applied Science and Engineering Research.* Vol *1*(2), 57-62.

[41] ASTM D 2867-09. 2009. *Standard test method for moisture in activated carbon: Designation.* ASTM International, West Conshohocken, USA.

[42] ASTM D2866-94. 1999. *Standard test method for total ash content of activated carbon.* ASTM International, West Conshohocken.

[43] Brunauer, S., Emmett, P. H., and Teller, E. 1938. Adsorption of gases in multimolecular layers. *Journal of the American Chemical Society.* Vol 60 (2), 309-319.

[44] Mozammel, H. M., Masahiro, O., and Bhattacharya, S. C. 2002. Activated charcoal from coconut shell using $ZnCl_2$ activation. *Biomass and Bioenergy.* Vol 22, 397-400.

[45] Yusup, S., Aminudin, A., Azizan, M. T., Abdullah, S. S., and Sabil, K. M. 2010. Improvement of rice husk and coconut shell properties for enhancement of gasification process, in: *Proceedings of Third International Symposium on Energy from Biomass and Waste*, Venice, Italy.

[46] Guo, J., and Lua, A. C. 2000. Textural characerization of activated carbon prepared from oil-palm stones with various imregnating agents. *Journal of Porous Materials.* Vol 7(4), 491-497.

BIOGRAPHICAL SKETCH

Iqbal Khalaf Erabee

Affiliation: Civil Engineering Dept., University Putra Malaysia, Civil Engineering Dept., University of Thi-Qar, Iraq

Education:

- Bachelor's degree (BSc) in Civil Engineering, University of AL-Mustansiriya, Baghdad, Iraq.
- Master degree (MSc) in fluid mechanics, University of AL-Mustansiriya, Baghdad/ Iraq.

Research and Professional Experience: In 2001, work as a lecturer at the Department of Civil Engineering, University of AL-Mustansiriya in Iraq for one year and then worked as a lecturer at the Department of Civil Engineering, University of Thi-Qar in Iraq and continuously until now. In 2014, enrolled as a Ph. D student at Universiti Putra Malaysia. Throughout PhD study, the researcher has got participation in worldwide conferences and eight published and submitted papers in ISI and Scopus journals.

Honors: Lecturer

Publications from the Last 3 Years:

[1] Erabee, I. K., Ahsan, A., Daud, N. N. N., Idrus, S., Shams, S., Din, M. F. M., & Rezania, S. (2017). Manufacture of Low-cost Activated Carbon Using Sago Palm Bark and Date Pits by Physiochemical Activation. *Bio Resources, 12*(1), 1916-1923.

[2] Erabee, I. K., Ahsan, A., Jose, B., Arunkumar, T., Sathyamurthy, R., Idrus, S., & Daud, N. N. (2017). Effects of electric potential, NaCl, pH and distance between electrodes on efficiency of electrolysis in landfill leachate treatment. *Journal of Environmental Science and Health, Part A*, 52 (8), 735-741.

[3] Erabee, I., Ahsan, A., Jose, B., Aziz, M. M. A., Ng, A., Idrus, S., & Daud, N. (2018). Adsorptive treatment of landfill leachate using activated carbon modified with three different methods. *KSCE Journal of Civil Engineering*, 2(4), 1-13.

Anne WM Ng

Affiliation: College of Engineering & Science, Victoria University, Melbourne, Australia

Education: B Civil Eng (Hons), PhD, Grad Dip Learning and Teaching

Research and Professional Experience: Dr Anne Ng is a Senior Lecturer in Water Engineering and has over 15 years of academic experience at various universities across Melbourne. Dr Ng's main research interest is Water Resources management, including management of Green Infrastructure and Urban Water Demand forecasting. She has published over 70 referred journal and conferences papers.

Areas of Expertise:

- Green Infrastructure
- Water resources management
- River water quality modelling.

Professional Appointments: Senior Lecturer – College of Engineering & Science, Victoria University, Melbourne, Australia

Publications from the Last 3 Years:

1. Imran, H. M.; Kala, J.; Ng, A. W. M.; Muthukumaran, S (2017) An evaluation of the performance of a WRF multi-physics ensemble for heatwave events over the city of Melbourne in southeast Australia, *Climate Dynamics*, 1-34
2. Erabee, I. K.; Ahsan, A.; Jose, B.; Aziz, M. M. A.; Ng, A. W. M.; Idrus, S.; Daud, N. N. N. (2017) Adsorptive treatment of landfill leachate using activated carbon modified with three different methods, *KSCE Journal of Civil Engineering*, 1-13

3. V. M. Jayasooriya, Ng, A. W. M., Muthukumaran, S., Perera, B. J. C. (2017) Optimal Sizing of Green Infrastructure Treatment Trains for Stormwater Management, *Water Resources Management* 30: 5407.
4. Sachindra, DA, Ng, A. W. M., Muthukumaran, S, Perera, B. J. C (2016), Impact of climate change on urban heat island effect and extreme temperatures: A case-study, *Quarterly Journal of the Royal Meteorological Society* 142(694):172-186.
5. Barua S, Ng A. W. M, Muthukumaran S, Roberts P, Perera, BJC (2015), Modeling water use in schools: A disaggregation approach, *Urban Water Journal* 1-7.
6. Jayasooriya, V. M. and Ng, A. W. M (2014), *Tools for modeling of Stormwater Management and Economics of Green Infrastructure Practices: a Review*, Vol 225 (8) pp 2055-2075.

About the Author

Monzur A. Imteaz is working as an Associate Professor in the Department of Civil and Construction Engineering at Swinburne University of Technology, Melbourne, Australia. He has obtained his Masters degree in 1994 on Water Resources Engineering from Asian Institute of Technology, Thailand. Then completed his PhD in 1997 on Lake Water Quality Modelling from Saitama University, Japan. After his PhD, he was working with the Institute of Water Modelling (Bangladesh). Later he has completed his postdoctoral research at The University of Queensland, Australia and subsequently he was working with several Australian government authorities. At Swinburne, he is teaching the subjects 'Urban Water Resources' and 'Integrated Water Design'. He has also been actively involved with various researches on sustainability, water recycling and modelling, developing decision support tools and rainfall forecasting using artificial neural networks. To date he is the author of one edited book, 15 book chapters, 113 journal and 93 conference papers. Also, he has successfully supervised 10 PhD students to completions.

In: Nutrients, Wastewater and Leachate
Editor: Amimul Ahsan
ISBN: 978-1-53613-949-5

Chapter 6

TREATMENT OF SCHEDULED WASTE LANDFILL LEACHATE: PERSPECTIVE IN MALAYSIA

Hasfalina Che Man[1,*], Nurul Hanira Mat Lazim[2], M. Rashid Mohd Yusof[3], Aida Isma Mohamad Idris[4], Luqman Chuah Abdullah[5] and Daniel I. Onwude[6]

[1,2,6]Department of Biological and Agricultural Engineering,
Universiti Putra Malaysia, Serdang, Selangor Darul Ehsan, Malaysia
[3]Air Resources Research Laboratory,
Malaysian-Japan International Institute of Technology,
University of Technology Malaysia, Kuala Lumpur, Malaysia
[4]Faculty of Engineering and the Built Environment,
Segi University, Petaling Jaya, elangor Darul Ehsan, Malaysia
[5]Department of Chemical and Environmental Engineering,
Faculty of Engineering, Universiti Putra Malaysia,
UPM Serdang, Selangor, Malaysia

[*] Corresponding Author Email: hasfalina@upm.edu.my.

ABSTRACT

In recent years, many advanced waste disposal methods have emerged. However, landfilling is still widely adopted in most countries, due to its low cost and simplicity. Nevertheless, the generation of a highly complex and polluted liquid leachate from landfill is one of major concerns as it threatens human health and environment. The constituents of leachate may vary, ranging from dissolved organic matter, inorganic macro compounds, heavy metals, ammonia-nitrogen (NH_3-N), bacteria and other microorganisms. The presence of different constituents in leachate had made it difficult to be treated, and different treatment approach is needed depending on the target pollutants. Despite the increasing number of documented scientific literature in various wastewater and leachate treatment, the literature showed gaps on understanding the characteristics and treatment methods of leachate from scheduled waste landfill (SWL) in Malaysia. Therefore, this chapter is significant in providing information on the characteristics of SWL leachate and technology available for treatment of SWL leachate.

There are various leachate treatments with different advantages and disadvantages. However, there is no known single technique that has been found to be applicable or highly effective in removing recalcitrant compounds, heavy metals, and NH_3-N from stabilized leachate. Sometimes, a combination of more than one treatment (biological, chemical and physical methods) is essential for the efficient treatment of leachate. Therefore, this chapter will also discuss the advantages and disadvantages of biological, chemical and physical methods of landfill leachate treatments as well as the combination of various treatment methods available in the literature. In addition, the qualitative and quantitative data on case study of scheduled waste leachate generated from a SWL treatment in Malaysia were also reported. Overall, SWL needs an effective and proper management to avoid environmental deterioration.

Keywords: scheduled waste landfill (SWL), hazardous waste, SWL leachate treatment, air stripping

1. INTRODUCTION

Landfilling is the most common method of waste disposal. This method, which involves the dumping of rubbish, garbage, and other forms of waste materials in a pre-designed site has been reported to generate highly complex and polluted liquid leachates which are harmful to human health and the environment (Di Iaconi et al., 2006; Renou et al., 2008; Tizaoui et al., 2007). These generated leachates are often small quantity, but with extremely hazardous contents (El-Fadel et al., 2002; Tizaoui et al., 2007). According to Di Iaconi et al. (2006), the high concentration of pollutants in landfill leachates lead to the contamination of ground water, which may affect both human and aquatic life (Tizaoui et al., 2007).

Several treatment technologies such as biological and physiochemical methods have been employed in removing the bulk of pollutants from leache before discharging into natural waters (Renou et al., 2008). However, there are limited studies on the characteristics and treatment methods of leachate from scheduled waste landfill (SWL) in Malaysia. The characterization of leachate, ammonia removal and treatment of scheduled wastes landfill in Malaysia is not well-documented. The amount of scheduled waste production is expected to rise each year. The fundamental aspect that needs to be considered in designing a sustainable scheduled waste management system is the availability of information on the characteristics of waste generated. However, such data on the Malaysian scheduled waste is limited. The purpose of this chapter is to introduce the reader to the field of scheduled waste landfill. More topics were elaborated in this chapter including the technology and management of SWL in Malaysia. A case study was included.

1.1. Scheduled Wastes Landfill in Malaysia

Generally, hazardous waste in Malaysia can be divided into 3 categories: general scheduled, clinical and radioactive wastes (Ujang, 1997). According to the recent overview by Aja et al. (2016), scheduled wastes are the categories of waste listed in the First Schedule of the Environmental Quality (Scheduled Wastes) Regulations 2005. Environmentally hazardous substance (EHS), under the Malaysian Environmental Quality Act (EQA) 1974, is defined as any natural or artificial substances including any raw material, whether in a solid, semi-solid, or liquid form, or in the form of gas or vapor, or in a mixture of at least two of these substances, or any living organism intended for any environmental protection, conservation, and control activity, which can cause pollution (DOE, 2012). There are currently 3839 items in the EHS reference list (DOE Online). Recently, there are 77 categories defined in the First Scheduled Waste of the Environmental Quality in Malaysia as EHS, which are classified into five groups as shown in Table 1.

The hazardous wastes in the five groups are from different sources such as industrial sector, agricultural sector, health sector, and households. Industrial waste poses potential serious hazard to the environment as most industrial processes employ chemical or chemically produced materials. All the details of scheduled wastes types can be found in the literature (Aja et al., 2016).

Table 1. Division of Scheduled Waste based on Environmental Quality (Scheduled Wastes) Regulation 2005 (Aja et al., 2016)

Waste Code	Description	Types of Scheduled Waste
SW 1	Metal and metal-bearing wastes	10
SW 2	Wastes containing principally inorganic constituents which may contain metals and organic materials	7
SW 3	Wastes containing principally organic constituents which may contain metals and inorganic materials	27
SW 4	Wastes which may contain either inorganic or organic constituents	32
SW 5	Other wastes	1

It has been reported that scheduled wastes in Malaysia has increased from 2,541,762.34 metric tonnes in 2014 to 2,918,478.34 metric tonnes in 2015, with the main categories of scheduled waste generated originated from dross, slag, clinker, ash, heavy metal sludge, and gypsum (Malaysia Environmental Quality Report, 2015). Typical scheduled waste generated, and its quantity is presented in Table 2 and scheduled wastes managed under special management is shown in Table 3. These waste streams were mostly produced by coal-fired power plant (44.71%) and heavy metal sludges from drinking water treatment facilities (41.51%).

Table 2. Quantity of scheduled waste generated by industry, Malaysia 2015

No.	Type of Industry	Quantity of Waste	
		MT/Year	Percentage (%)
1	Power Plant	829,827.06	28.43
2	Water Treatment Plant	648,271.37	22.21
3	Recovery Facilities	511,643.12	17.53
4	Chemical Industry	191,199.00	6.55
5	Metal Refinery	169,057.30	5.79
6	Electric And Electronic	138,085.75	4.73
7	Paper	118,878.67	4.07
8	Vehicle	44,242.67	1.52
9	Others	35,366.94	1.23
10	Excavation Non Metal	33,907.16	1.16
11	Health Care Services	25,523.33	0.87
12	Printing	23,677.92	0.81
13	Workshop	23,364.27	0.80
14	Rubber Base	22,527.82	0.77
15	Petroleum Refinery	20,908.57	0.72
16	Metal Finishing and Coating	15,879.90	0.54
17	Food & Drink	8,953.58	0.31
18	Metal Fabrication	8,760.25	0.30
19	Edible Oil Refinery	7,020.69	0.24
20	Mining	5,868.01	0.20
21	Services	5,728.52	0.20
22	Plastic	5,098.91	0.17
23	Quarry	4,087.28	0.14
24	Warehouse	3,751.56	0.13
25	Agriculture	3,330.23	0.11
26	Machinery	3,302.68	0.11
27	Palm Oil Mill	2,923.73	0.10
28	Others Manufacturing	2,680.55	0.09
29	Wood Base	1,128.11	0.04
30	Cement	1,042.93	0.04
31	Textiles	1,040.38	0.04
32	Sanitary Landfill	516.91	0.02
33	Rubber Factory	473.89	0.02
34	Sports Equipment and Games	244.97	0.01
35	Office Supplies and Stationery	164.29	0.01
	Total	2,918,478.34	100.00

Source: *Malaysia Environmental Quality Report*, 2015

These waste streams generate high number of leachates which could have great environmental and health impact if not properly treatment. Therefore, the efficient treatment of SWL in Malaysia becomes indispensable.

Table 3. Generated Scheduled Wastes Managed Under Special Management 2015

No.	Waste Category	Waste Code	Source	Metric Tonnes	Percentage (%)
1	Heavy metal sludge	SW 204	Drinking water treatment plant	645,498.62	41.51
			Industry	83,028.01	5.34
				14,980.55	0.96
2	Fly ash and bottom ash	SW 104	Coal-fired power plant	695,299.87	44.71
			Industry	24,738.34	1.59
3	Gypsum	SW 205	Industry	73,189.91	4.71
				344.80	0.02
4	Waste containing formaldehyde, resin, discarded epoxy powder	SW 320, 325, 418	Industry	7,053.66	0.45
5	Discarded pharmaceutical product, discarded product	SW 405, 429	Industry	30.59	0.00
6	Ash of paper sludge	SW 406	Industry	2,436.35	0.16
7	Spent mixed oil	SW 421	Industry	6,628.95	0.43
8	Mineral sludge	SW 427	Industry	1,773.43	0.11
			Total	1,555,003.07	100.00

Source: *Malaysia Environmental Quality Report*, 2015

2. Malaysian Legislation and Regulations on Scheduled Wastes Landfill Leachate Management

In Malaysia, toxic and hazardous wastes or scheduled waste are regulated by Environmental Quality (Scheduled Wastes) Regulation 1989 under Environmental Quality Act, 1974, which has been replaced by the

Environmental Quality (Scheduled Wastes) Regulations, 2005. Table 4 gives a detailed account of scheduled wastes management history, milestones, laws and regulations since it had been commissioned. Since then, numerous regulations regarding hazardous waste have been introduced, including the definition and classification of hazardous waste, collection, storage, transportation, treatment and disposal facilities.

Table 4. History of Malaysia Laws and Regulations on Scheduled Waste

Year	Description	Notes
1960s	Malaysia placed great emphasis on industrial growth as developing country	RMK-1 (1966-1970) (Industrialization)
1974	The Environmental Quality Act (EQA) 1974 was gazetted and come into force on 15 April 1974	-
1979s	DOE initiated guidelines for handling and disposal of toxic and hazardous wastes	Event of uncontrolled and indiscriminate waste disposal
1984	DOE began drafting regulation for management of toxic and hazardous waste in Malaysia	-
1989	The final draft of proposed scheduled waste regulation was gazetted on 27 April 1989 and came into force on 1 May 1989	Environmental Quality (Scheduled Wastes) Regulations 1989
1992	On 8 January 1992, government awarded to consortium named Kualiti Alam Sdn Bhd to build integrated scheduled waste facilities in Malaysia	-
1993	On 8 October 1993, Malaysia ratified Basel Convention on the Control of Transboundary Movement of Hazardous Wastes and Their Disposal	-
1998	Kualiti Alam Sdn Bhd facilities was completed in Jun 1998 and official operated as the first integrated treatment and disposal facility in this region	-
2005	Over the 15 years, DOE revised scheduled waste regulation and new regulations was proposed and came into force	Environmental Quality (Scheduled Wastes) Regulations 2005
>2015	11th Malaysia National Plan -way forward for holistic scheduled waste management in Malaysia e.g., 'cradle to cradle concept', e-waste management, special waste management	-

Source: *Department of Environment*, 2015

Scheduled wastes are potentially harmful. The normal handling such as storage, packaging, transport and subsequent disposal methods that are usually used for solid wastes or non-hazardous industrial wastes (typically landfilling) are not appropriate for scheduled wastes as it presents potential risks to the human health and environment. Till today, most of scheduled wastes generated were disposed of at landfills owned by local authorities. Nevertheless, disposal at sanitary landfills is still not efficient, to avoid soil and groundwater contamination. Scheduled wastes have to be landfilled at secure landfill to meet the toxicity guidelines (TCLP) as proposed by the Department of Environment Malaysia. In general, characteristics of scheduled wastes are ignitable or inflammable, reactive, corrosive, toxic and infectious (Department of Environment Malaysia, 2015).

The waste generator of scheduled wastes may store, recover or treat the waste within their own premises. These activities do not need approval by the Department of Environment. A waste generator may store scheduled wastes generated by him for 180 days or less after its generation provided that the quantity of scheduled wastes accumulated on site shall not exceed 20 metric tonnes. However, they may apply to the Director General in writing to store more than 20 metric tonnes of scheduled wastes. The containers that are used to store scheduled wastes shall be clearly labelled with the date when the scheduled wastes are first generated including the name, address and telephone number of the waste generator. Land farming, incineration, disposal and off-site facilities for recovery, storage and treatment can only be carried out at prescribed premises licensed by the Department of Environment. On-site incineration of scheduled wastes is not encouraged. If it is deemed necessary, application for the installation of such incinerator must strictly adhere to the Guidelines On the Installation of Onsite Incinerator for the Disposal of Scheduled Wastes in Malaysia (published by the Department of Environment), including carrying out a detailed environmental impact assessment and display of the EIA report for public comments. Waste generators may apply for special management of scheduled wastes to have the scheduled wastes generated from their particular facility or process excluded from being treated, disposed of or recovered in premises or facilities other than at the prescribed premises or onsite treatment or recovery facilities, as stipulated under Regulation 7(1), Environmental Quality (Scheduled Wastes) Regulations 2005. Kualiti

Alam Sdn. Bhd. has been granted the exclusive right to operate scheduled waste management system including waste collection, transportation, treatment, and final disposal (Aja et al., 2016).

Generally, for final disposal, scheduled waste is sent to the licensed facilities (Kualiti Alam Sdn Bhd and Trienekens (Sarawak) Sdn Bhd). The types of wastes sent includes sludge containing one or several heavy metals, mixed wastes, dust/slag/dross or ash containing arsenic/mercury and spent inorganic acid. The wastes will go through several types of treatments, either incinerated, treated physically and chemically, solidified or disposed of in secured landfill depending on their characteristics. The most applied treatments for scheduled wastes which were sent to Kualiti Alam Sdn Bhd and Trienekens Sdn Bhd. are; 54.8% of the scheduled waste received were landfilled, followed by 34.6% of incineration, 7.7% of solidification and 2.9% of waste underwent physical and chemical treatment.

Table 5. Facilities Handling Scheduled Wastes, 2015

No.	Facility	Metric Tonnes (MT)	Percentage (%)
1	Special Waste Management	1,555,003.07	53.28
2	Local Off-site Recovery Facilities	446,077.76	15.28
3	On-Site Treatment	445,867.97	15.28
4	Kualiti Alam Sdn Bhd	120,221.72	4.12
5	On-Site Storage	299,622.49	10.27
6	Off-site Clinical Waste Incinerators	25,523.33	0.87
7	Trienekens (Sarawak) Sdn Bhd	19,283.00	0.66
8	Foreign Facilities (Export)	6,879.00	0.24
	Total	2,918,478.34	100.00

Source: *Malaysia Environmental Quality Report*, 2015

According to the Malaysia Environmental Quality Report (2015), from the overall scheduled waste generated in 2015, 1,555,003.07 MT (53.24%) were granted conditional approval to be managed under special management. In addition, for the scheduled waste treated locally, a total of 120,221.72 MT (4.12%) were treated and disposed for final disposal at Kualiti Alam Sdn. Bhd, Trienekens (Sarawak) Sdn. Bhd. (19,283.00 MT) and 25,523.32 MT (0.87%) of clinical wastes were incinerated at licensed off-site facilities (Table 5).

Furthermore, a total 445,867.97 MT (15.28%) of scheduled waste were treated on-site, while the remaining 299,622.49 MT (10.27%) were stored onsite at waste generators' premises as described in Table 5. Two (2) land farms and fifteen (15) on-site waste incinerators have been licensed by Department of Environment Malaysia to allow for on-site treatment and incineration, respectively.

Table 6. Numbers of Off-site Recovery Facilities for Scheduled Wastes in Malaysia

Waste Category	Recovery Facility
Electronic and Electrical Wastes	140
Dross/ Slag / Clinker	64
Oil/ Mineral Sludge/ Spent Coolant	51
Heavy Metal Sludge	49
Used Container	48
Paper & Plastic	41
Spent Solvent	39
Acid/ Alkaline	34
Phenol/ Adhesive/ Resin	32
Mixed Wastes	27
Ink/Paint	23
Catalyst	19
Photographic	12
Battery	11
Rubber sludge	10
Gypsum	6

Source: *Malaysia Environmental Quality Report*, 2015

Department of Environment Malaysia had given the license to the facilities depending on the waste categories which are electronic and electrical wastes (140 facilities) followed by dross/ash/slag (64 facilities), oil/mineral sludge/spent coolant (51 facilities), heavy metal sludge (49 facilities), used container (48 facilities), paper and plastic (41 facilities), spent solvent (39 facilities), acid and alkali (34 facilities), phenol/

adhesive/resin (32 facilities), mixed wastes (27 facilities), ink/ paint (23 facilities), catalyst (19 facilities) and the remaining 4 wastes categories totaling of 39 facilities as shown in Table 6.

3. CHARACTERISTICS OF SCHEDULED WASTES LANDFILL LEACHATE

Table 7 demonstrates three classification of leachate according to the landfill age: young leachate, medium or middle age leachate and stabilized or old leachate. Occasionally, young leachate and medium age leachate are merged as one category: young leachate below 5 years. The characteristics of leachate under this category consists of high concentration of biodegradable organic compounds such as volatile fatty acids (VFAs), high concentration of COD (6,000 to 60,000 mg L^{-1}), high BOD (4,000 – 13,000 mg L^{-1}), a high ratio of BOD/COD ranging from 0.4 to 0.7 and high concentration of NH_4^+-N (2,000 – 5,000 mg L^{-1}) (Welander et al. 1998; Wang et al. 2003; Kurniawan, 2011).

Table 7. Landfill leachate classification (Alvarez-Vazquez et al., 2004)

	Young	**Medium Age/Intermediate**	**Stabilized/Old**
Age of landfill (years)	< 1	1-5	> 5
pH	< 6.5	6.5-7.5	> 7.5
COD (mg L^{-1})	> 15000	3000-15000	< 3000
BOD_5/COD ratio	> 0.5	0.1-0.5	< 0.1
NH_4^+-N (mg L^{-1})	< 400	NA	> 400
TOC/COD	< 0.3	0.3-0.5	> 0.5
Kjeldahl nitrogen (g L^{-1})	0.1-2	NA	NA
Heavy metals (mg L^{-1})	> 2	< 2	< 2

As landfill age increases, the degradation of organic materials into methane and carbon dioxide (CO_2) by microorganisms occurred. During this process, CO_2 is reduce with hydrogen (H_2), thus increases the pH to over 7.0. At this stage, the leachate organic compounds are not biodegradable anymore and turn to stabilized leachate (Mårtensson et al, 1999). This phase may be prolonged to 50 years or more (Gurijala and Suflita, 1993). In old landfill (more than 10 years), refractory compounds dominates the organic fraction of the leachate (Welander et al. 1998; Wang et al. 2003).

Table 8. Permissible limit standards for SWL leachate discharged

Parameters*	**Limiting Concentrations**
Temperature	40
pH value	5.5-9.0
BOD_5 at 20 °C	50
Suspended Solids	100
Mercury	0.05
Cadmium	0.02
Chromium, Hexavalent	0.05
Chromium, Trivalent	1.0
Arsenic	0.10
Cyanide	0.10
Lead	0.5
Copper	1.0
Manganese	1.0
Nickel	1.0
Tin	1.0
Selenium	0.5
Barium	2.0
Fluoride	5.0
Formaldehyde	2.0
Phenol	1.0
Free Chlorine	2.0
Sulphide	0.5
Oil and Grease	10
Ammoniacal- nitrogen (NH_3-N)	20
Colour	200

*Note: All measurement units of parameters are in mg L^{-1}, except for temperature (°C), pH value (unitless) and colour (ADMI).

Source: Environmental Quality (Industrial Effluent or Mixed Effluent) Regulations 2009, Fifth Schedule (Standard B).

The stabilized leachate comprises of high concentration of COD (5,000 – 20,000 mg/L), high concentration of NH_4^+-N (2,500 – 5,000 mg/L), and low BOD/COD ratio less than 0.1 (Kurniawan, 2011). In Malaysia, scheduled waste landfill leachate is categorised under Environmental Quality (Industrial Effluent or Mixed Effluent) Regulations 2009, Fifth Schedule (Standard B). Table 8 gives the permissible limit standards for scheduled waste landfill leachate discharged in Malaysia.

SWL leachate is rarely studied due to its complex characteristics and hazard-containing materials compared to other leachate types (Sani et al., 2014; Hanira et al., 2015). Table 9 shows the comparison between characteristics of SWL and municipal solid waste in Malaysia. The comparisons include; AKSL = Anaerobic Kulim Sanitary Landfill, SPBSL = Semi-aerobic Pulau Burung Sanitary Landfill and KSSL = Kuala Sepetang Sanitary Landfill. The characteristics of SWL leachate are different from other landfill leachate. Sani (2014) reported the average NH_3-N concentration for SWL leachate in 2013 was 1906 ± 590 mgL^{-1}, with the range of 1090 to 2945 mgL^{-1} varied from January to December. In addition, SWL leachate comprised of high metal content such as Boron (B), Arsenic (As), Copper (Cu) and Nickel (Ni), which are derived from industrial wastes, with Boron and Arsenic seeming to be the distinctive component of SWL leachate as other leachates contain low traces of these metals. SWL leachate has higher pH, COD, BOD_5 and NH_3-N compared to other landfill leachates. The BOD_5/COD ratio for SWL leachate and other leachates was in the range of 0.57 to 0.62, and 0.10 to 0.20, respectively. Furthermore, high pH level (10.02) of SWL leachate (Hanira et al., 2015) is also not a suitable condition for bacteria to culture as the optimal nitrification rates occur at pH values of 7.5 to 8.0 (Kutty et al., 2011). Due to these reasons, direct biological treatment is inappropriate for SWL leachate.

According to Zhang (2015), the treatment of landfill leachates is difficult due to (1) high concentration and complex components of organic pollutants, (2) disproportion of nutrient elements (Carbon/Nitrogen/

Phosphorus), (3) high concentration of a certain amount of metal ions and ammonia ions, and (4) great variation of water quality of leachate, and biodegradability of leachate decreases by the burying time.

Table 9. Physico-chemical characterization of raw SWL leachate and comparison with other landfill leachate in Malaysia

Parameter	Units	SWL Leachate (Sani et al., 2014)	SWL Leachate (Hanira et al., 2015)	AKSL leachate (Aziz et al., 2015)	SPBSL leachate (Aziz et al., 2015)	KSSL leachate (Aziz et al., 2015)
pH Value	-	9.92	10.1	8.02	8.36	8.66
Biochemical Oxygen Demand (BOD_5)	mgL^{-1}	2873	2275	285	181	257
Chemical Oxygen Demand	mgL^{-1}	4633	4020	1295	1819	1456
BOD/COD ratio	-	0.62	0.57	0.20	0.10	0.19
Ammoniacal-Nitrogen	mgL^{-1}	2403	2180	562	1627	564
Total Iron	mgL^{-1}	0.4	ND	3.82	4.9	3.43

3.1. High Concentration and Complex Components of Organic Pollutants

The characteristics of landfill leachate generated depend on the types of wastes buried in landfill. Some compounds such as phenolics, amine substance, and heterocyclic compounds are difficult to be degraded (Zhang, 2015). Whereas, some compounds such as heavy metals and NH_3-N are carcinogenic and toxic (Białowiec, 2011). In general, municipal solid waste and sanitary landfills have higher concentration of nitrogenous compounds which originates from food waste (protein). In fact, they have low concentration of toxic compounds such as heavy metals because they are not industrial wastes (Zhang, 2015). This condition is suitable for the growth of microorganisms and direct biological treatment is applicable for this leachate type since no toxic compound present in the leachate that might restrict the microorganism growth.

3.2. Disproportion of Nutrient Elements (Carbon/Nitrogen/ Phosphorus)

Most solid waste and sanitary landfill leachate are generated from domestic waste, specifically kitchen garbage, which contains high percentage of organics material. In contrast, the inorganic materials decrease rapidly, which consequently contribute to higher COD, BOD and NH_3-N. Concentration of NH_3-N is inversely proportional to the Phosphorus (P) content. High concentration of NH_3-N lowers the P content in leachate. This condition may disrupt the growth of microorganisms, which results in the difficulty of biodegradation (Zhang, 2015). As an example, SWL leachate has high NH_3-N concentration which are contributed by industrial wastes such as dross, slag, clinker, ash, heavy metal sludge, and gypsum. High NH_3-N concentration leads to low P content which makes the biodegradation problematic.

3.3. High Concentration of a Certain Amount of Metal Ions and Ammonia Ions

Landfill leachate constituents vary and may consist of certain amount of metal ions, for examples Mangan (Mn), Nickel (Ni), Lead (Pb) and others. Landfill leachate which is generated from domestic wastes generally has low metal ions content. However, for landfill leachate generated from mixed landfill of industry and domestic wastes, it may have higher density of metal ions, such as Cu, Zinc (Zn), Pb, Ni and Mercury (Hg). Furthermore, another factor which makes leachate hard to be treated is the high concentration and dynamic change of NH_3-N in leachate. It is noted that an ammonia concentration about 1000 mgL^{-1} is toxic to microorganisms (Shiskoswski and Mavinic, 1998; Wiszniowski et al., 2007) and higher NH_3-N concentration of 1500–3000 mg L^{-1} inhibits the biodegradation process by nitrite oxidizing bacteria (NOB) and ammonia oxidizing bacteria (AOB) (Liu et al., 2012; Sani et al., 2014). The release of leachate containing NH_3-N into surface waters may enhance the growth of algae and aquatic plants, cause eutrophication in lakes, threaten the life of aquatic animals and deteriorate the balance of ecosystem (Clement and Merlin, 1995; Jokela et al., 2002; Cai et al., 2013).

When the landfill leachates enter the stage of producing methane, the concentration of NH_3-N will continue to increase. As NH_3-N reaches its highest density, it will last for a long period of time. NH_3-N concentration in leachate might vary, ranging from 200 mgL^{-1} at the beginning to 1200 mgL^{-1}, and even up to 1500 mgL^{-1} (Zhang, 2015). For SWL leachate, it is reported that the average of NH_3-N concentrations is 2000 mgL^{-1} (Sani, 2016). This NH_3-N value is the highest compared to other types of landfill leachate in Malaysia, namely Pulau Burung Landfill Site with 1380 mgL^{-1} (Umar et al., 2010), Kulim Landfill Site with 210 mgL^{-1} (Zainol et al., 2012); Ampang Jajar Landfill Site with 566 mgL^{-1} (Umar et al., 2010), Kuala Sepetang Landfill Site with 857 mgL^{-1} (Zainol et al., 2012), Sahom Landfill Site with 1113 mgL^{-1} (Azmi et al., 2014), Air Hitam Leachate Site with 0.28 mgL^{-1} (Salleh & Hamid, 2013), Sungai Sedu Landfill Site with 29 mgL^{-1} (Hamzah et al., 2014) and Ampar Tenang Landfill Site with 14 mgL^{-1} (Yusof et al., 2013). Additionally, NH_3-N concentration changes dynamically with burying time. In fact, high concentration of NH_3-N leads to the low ratio of C/N. As low ratio of C/N restrains the effectiveness of treatment by biological method, the high NH_3-N content in leache requires the process of nitrogen removal prior biological treatment. Since NH_3-N is one of the compounds that cause toxicity to most organisms, there is need for proper treatment and removal from landfill leachate (Clement and Merlin, 1995; Tam and Wong, 1996; Jokela et al., 2002).

3.4. Great Variation of Water Quality of Leachate, and Biodegradability of Leachate Decreases by the Burying Time

In landfill leachate, microorganisms normally exist in large amount, which are important for waste biodegradation. Suitable temperature and humidity in landfill accelerates the growth and reproduction of the microorganisms. Organic materials in leachate can be divided into three groups which are equimolecular griseofulvin acid substances, humic polymer carbohydrate and low molecular weight fatty acid. At the early stage of burying, the colour of leachate is black and 90% of the dissolved organics in leachate is volatile fatty acid and griseofulvin acid. The BOD concentration is high in leachate, which relatively high biodegradability. Overtime, the volatile fatty acid reduced slowly. In contrast, the ratio of the griseofulvin substance increased. Hence, the changes of organic materials in leachate results in the decline of BOD/COD, which means the reduction in the biodegradability of leachate. At the end of burying period, the colour of leachate turn into brown, while the NH_3-N concentration increase rapidly, adding the difficulty for the leachate treatment (Zhang, 2015).

4. Treatment of Landfill Leachate

There are various treatments of landfill leachate including biological and physiochemical treatments as summarized in subtopic 4.1 and 4.2, with different advantages and drawbacks of each treatments. Biological treatment is preferred due to its low cost and simplicity for landfill leachate pollutant removal (Yang and Zhou, 2008). Biological treatment is quite interesting as it is more environmentally friendly. However, this technique is inappropriate for leachate or wastewaters that contained very high concentration of NH_3-N such as SWL leachate. Numerous studies have reported that biological methods are only efficient for young leachates but not for old/stabilized leachate. This is due to a high fraction of nonbiodegradable, toxic chemicals and high ammonia concentrations in old/stabilized leachate (Renou et al., 2008). To date, there is no single technique that is applicable or highly effective in removing recalcitrant compounds, heavy metals, and NH_3-N from stabilized leachate and/or SWL leachate. Hence, a combination of more than one treatment (biological, chemical and physical methods) is essential for the efficient treatment of leachate (Kurniawan et al. 2006; Kurniawan, 2011; Liu et al., 2013; Zhang et al. 2012).

4.1. Biological Treatment

Biological treatment is the process in which wastewater or leachate is treated with anaerobic or aerobic bacteria in order to remove or reduce organic contaminants for instance organic matter, ammonia and nitrates. It is based on the suspended-growth biomass, which includes conventional activated sludge processes, stabilization ponds, aerated lagoons, membrane bioreactor, trickling filters, Moving-Bed Biofilm Reactor (MBBR) (or Suspended-Carrier Biofilm Reactor (SCBR) or Fluidized Bed Reactor), rotating biological contactors (RBC) and sequencing batch reactors.

The biological treatment can be categorised as aerobic and anaerobic treatment (Metcalf and Eddy, 1991). Generally, it involves placing a waste stream in direct contact with a mixed culture of microorganisms and comprises of monitoring the environment essential for optimum growth of microorganisms is used for treating leachate (Wiszniowski et al., 2007). The microorganisms will help biodegradation process by degrading the organic compounds to carbon dioxide (CO_2) and sludge under aerobic conditions to biogas (a mixture comprising of CO_2 and CH_4) under anaerobic conditions (Lema et al., 1988). The microorganisms will consume the organic matter present in the waste stream and leave the inorganic compounds unaffected. Biological treatment will have no effect on inorganic compounds. It is very effective in the removal of organic and nitrogenous matter from immature leachates when the BOD/COD ratio has a high value (>0.5) (Metcalf and Eddy, 1991; Zheng et al., 2015; Murunga et al., 2016). Low BOD_5/COD ratio (<0.4) was generally unfavourable for biological treatment process. However, in long term, the major presence of refractory compounds tends to limit the process's efficiency (Renou et al., 2008).

Meanwhile, anaerobic digestion technique comprises the decomposition of organic and inorganic matter without oxygen. Compared to aerobic treatment process, anaerobic digestion treatment conserves energy and low solid production, but suffers from low reactions rates (Berrueta and Castrillón, 1992). The anaerobic digestion treatment includes up-flow anaerobic sludge blanket (UASB) reactor, anaerobic filter, and hybrid bed filter. Table 10 summarize the methods, advantageous and disadvantageous of biological treatment.

Table 10. Summary of the advantages and disadvantages of biological methods for the treatment of landfill leachate

Method	Advantages	Disadvantages	References
Activated sludge	• Adapted to any size of community (except for very small ones) • Good elimination of all the pollutant parameters (Suspended solids, COD, BOD, and N) • Partially stabilizes sludge	• Relatively high capital costs • High energy consumption • Requires skilled personnel and regular monitoring • Sensitivity to hydraulic overloads	Lema et al. (1988) Hoilijoki et al. (2000) Blauvelt (2009)
Anaerobic Filter	• Suitable for treating some soluble wastes without effluent or solids recycle • Did not require specific temperature • Low sludge production	• Additional cost for support media	Young and McCarty (1969) Wang and Banks (2007)
Aerated lagoons	• This method can remove 60% to 90% of BOD • Adapted to wastewater and landfill leachate treatment due to large adsorptive surfaces on sediments, high productivity plant, aerobic anaerobic interfaces, plants and roots, and various populations of active microorganisms • Need little specialized skill to run the system • Low-cost method for installation, operation and maintenance • Its effectiveness for removing pathogens, organic and inorganic matters.	• Since this type of pond are not well mixed, there will be some solids still settle to the bottom and undergo anaerobic decomposition even the pond had been provided with extra aeration • Cannot achieve high effluent quality standards. • Temperature may affect the treatment processes by aerated lagoons as extreme temperatures (too low or high) affects microbial activity • Not suitable for leachate treatment as long retention time needed (approximately 3 to 20 days) for the biodegradation of organic matters.	Qasim (1994) Maynard et al. (1999) Renou et al. (2008)

Table 10. (Continued)

Method	Advantages	Disadvantages	References
Trickling filters	• Widely used for its low-cost filter media made by bed rock, • slag, or plastic for nitrification process • Can achieve high treatment efficiency in the removal of turbidity, suspended solids, biochemical oxygen demand and ammonium.	• High occurrence of clogging • Requirement of long rest period • Relatively low loading that could be used	Metcalf and Eddy (1991) Jokela et al. (2002) Aluko and Sridhar (2013)
Moving-Bed Biofilm Reactor (MBBR) or Suspended-Carrier Biofilm Reactor (SCBR) or Fluidized Bed Reactor	• Higher biomass concentrations process • Lower sensitivity to toxic compounds • Short period of sludge-settling • Higher organic and ammonia removals in a single process	• High operating cost due to aeration process.	Horan et al. (1997) Loukidou and Zoboulis (2001) Renou et al. (2008)
Sequencing Batch Reactors (SBR)	• Minimum operator interaction required and easy to operate. • Flexibility to treat high degree of variability in leachate quantity and quality • Efficient in removing organic matters and solids.	• High energy consumption • High sludge production • Could not treat leachate with low biodegradability	Olsson and Lo (1997) Kennedy and Lentz (2000) Kurniawan (2011) Liu (2013)
Up-flow Anaerobic Sludge Blanket (UASB) reactor	• Can achieve high treatment efficiency and a short hydraulic retention time • Leachate can be treated on-site at low temperature • Minimize the need for heating the leachate prior to treatment processes, which may provide a cost-effective treatment option.	• May produce toxic substances at low temperature	Kettunen et al. (1996) Lin and Chang (2000) Abbas et al. (2009)

Method	Advantages	Disadvantages	References
Rotating Biological Contactors (RBC)	• Low energy consumption for aeration • Less space occupation • Simple operation requiring less maintenance and monitoring than activated sludge technique • Good settling characteristics of the sludge • Generally adapted to small communities • Lower sensitivity to load variations and toxins than activated sludge • Resistance to cold (the disks are always protected by hoods or a small chamber).	• Performance is generally lower than with an activated sludge technique • High capital costs (can be greater by about 20% compared to activated sludge) • High strength leachate treatment via RBC is limited due to clogging problems.	Metcalf and Eddy (1991) Kurniawan (2011)

4.2. Chemical and Physical Treatment

Currently, chemically enhanced leachate treatment had created substantial interest, especially for leachate that cannot be treated by conventional biological treatment strategies. Biological treatment method may not always be suitable for leachate treatment particularly for certain circumstances such as high concentration of bio-toxicity in leachate, intermittent waste flows and leachate containing non-biodegradable impurities. It was found that biological treatment performed relatively reasonable (Table 10). However, it is difficult to achieve high and reliable ammonia and COD removal efficiencies alone. These methods are used in conjunction with biological processes in order to achieve permissible limit standards. Therefore, the combination of biological and chemical-physical techniques is generally practiced. Chemical and physical treatments have also been used widely for treating most industrial effluents, wastewaters and leachate. These treatment methods are usually applied for stabilized landfill leachate that consists of high concentration of bio-toxicity and lack degradable organic matters. Chemical-physical treatment is a non-biological method, which is frequently used as a pre-treatment and a post-treatment method in leachate treatment (Kjeldsen et al., 2002; Kurniawan et al., 2006; Renou et al., 2008). The treatment process involves the addition of chemicals into the waste stream to remove contaminants. The most common chemical and physical treatment processes of leachate are coagulation, flocculation, precipitation, sedimentation, carbon adsorption, chemical oxidation, reverse osmosis, ion exchange, electrochemical treatment and air/ammonia stripping. Table 11 shows the summary of the advantages and disadvantages of chemical and physical methods for the treatment of landfill leachate.

Table 11. Summary of the advantages and disadvantages of chemical and physical methods for the treatment of landfill leachate

Method	Advantages	Disadvantages	References
Coagulation-flocculation, sedimentation	• Capability of the treatment • Simplicity of the process • Inexpensive equipment employed	• High operational cost due to high chemical consumption • Sensitivity of the process to pH • High sludge production and subsequent disposal may be a problem	Metcalf and Eddy (1991) Qasim (1994) Kurniawan et al. (2006) Kurniawan (2011) Sani et al. (2014) Hanira et al. (2015)
Chemical oxidation and AOP (advanced oxidation process)	• Simple technology • No mass transfer limitation (homogeneous nature) • Both iron and hydrogen peroxide are cheap and non-toxic	• Need a lot of electrical energy, which results in higher treatment costs • High oxidant doses are required, thus adding the treatment cost • Some intermediate oxidation products may raise toxicity	Metcalf and Eddy (1991) Lopez et al. (2004) Wiznowski et al. (2006)
Chemical precipitation	• Capability of the treatment • Low-cost equipment and installation • Simplicity of the process	• High dose of precipitant required • Sensitivity of the process to pH • The production of sludge from the treatment • Requires further disposal due to sludge generation	Ozturk et al. (2003) Kurniawan et al. (2006) Kurniawan (2011) Liu (2013)

Method	Advantages	Disadvantages	References
	• For certain chemical such as struvite, the sludge produced after treatment may be utilized as a nitrogen fertilizer if the leachate does not contain any heavy metals		
Activated carbon adsorption	• High efficiency in VOC removal • Simple technology employed • Easy to handle and maintain • Do not require electrical energy • May improve the effluent quality by removing odours and dissolved gases and even can be effective against certain microorganisms	• Need frequent regeneration of activated carbon column and the high cost of GAC, which made this method less applicable for landfill leachate treatment in most developing countries • The tendency to blockage caused by dust, mixing of component that may lead to early malfunction, risk of spontaneous combustion in the bed such as ketones and turpentines and polymerisation risk for unsaturated hydrocarbons on the activated carbon, which may cause blockages • Not very effective in removing ammonia/ ammonium	Metcalf and Eddy (1991) Aziz et al. (2004) Kurniawan et al. (2006)
Ion Exchange	• Effective in removing the traces of metal impurities from landfill leachate • Easy installation, produce immediate results, minimal maintenance of operation and it is capable of meeting the permissible limit standards	• Prior ion exchange treatment, an appropriate pre-treatment is crucial for the removal of suspended solids. • The leachate must be treated with biological treatment first before ion exchange treatment was applied. • Not economically appealing due to high operational cost. • Not very effective for low pH leachate, complex mixtures of metals, resin regeneration and resin fouling.	Metcalf and Eddy (1991) Fettig et al. (1996) Lin and Wu (1996) Kurniawan et al. (2006) Kurniawan (2011)

Table 11. (Continued)

Method	Advantages	Disadvantages	References
Electro-chemical Treatment	• An option method to breakdown recalcitrant organic	• It consume high energy • More expensive compared to other treatment methods, thus, it is rarely applicable for landfill leachate treatment	Ali et al. (2004) Moraes & Bertazzoli. (2005)
	compounds in landfill leachate • May remove COD up to 73% with initial concentrations of 1855 mg/L • Not very effective for NH_3–N removal.	• Efficiency depends on type of electrode and substrate	
Reverse Osmosis	• Efficient for the removal of COD in leachate compared to other method • Reverse osmosis might remove total dissolved solids • up to 99%, NH3-N up to 95-97% and removal of COD up to 85- 95% • Can be used to remove heavy metals, suspended/colloidal materials and dissolved solids from landfill leachate • High fluxes and ability to operate within wide range of temperature and pH	• Pre-treatment is crucial in order to make sure reverse osmosis process is at most efficient • Filtration of the wastewater • is needed as reverse osmosis unit can be damaged by colloidal matter in the feed stream. Need to remove manganese and iron first to reduce scaling potential and maintained the pH in 4.0 to 7.5 ranges • High energy consumption • A low retention of small molecules that pass through the membrane • Membrane fouling An undesirable deposition of suspended or dissolved substances on the external surface of membrane	Metcalf and Eddy (1991) Qasim (1994) Choo and Lee (1996)

Method	Advantages	Disadvantages	References
Ammonia Stripping/ Air Stripping	• Simple technology • Very high VOC removal efficiencies (more than 90%) • It is possible to meet the NH_3-N discharge standards • More economic than other treatments for instance reverse osmosis or nanofiltration in terms of operation cost.	• Not cost-effective as need to increase the pH prior the treatment for increasing the removal rate of ammonia in leachate involves the use of chemicals • Coagulant might formed sludge or residue such as calcium carbonate scaling on the tower • Good for only compounds with high Henry's law constant • Off gas need further treatment • Additional cost is needed for pH adjustment of the treated effluent prior to discharge.	Metcalf and Eddy (1991) Cheung et al. (1997) Marttinen et al. (2002) Blauvelt (2009) El-Gohary et al. (2013) Ferraz et al. (2013) Hossini et al. (2015)

5. Treatment of Scheduled Waste Landfill Leachate

Applications of numerous techniques for landfill leachate have been reported by several authors (section 4). These include biological, chemical and physical treatments. However, with the complex characteristics of SWL leachate, the application of biological treatment is not an option. Direct biological treatment resulted in low removals due to high concentrations of NH_3-N in leachate, as it is toxic and inhibits the treatment process (El-Gohary et al., 2013). Pre-treatment prior to the biological treatment is required to reduce the concentration of NH_3-N (Li and Zhao, 1999; El-Gohary et al., 2013). Therefore, chemical and physical treatments are considered for SWL leachate treatment. The commonly adopted chemical and physical treatment for leachate had been discussed in detail in subtopic 5.1, focusing on chemical precipitation and ammonia stripping (subtopic 5.2).

5.1. Chemical Precipitation

Chemical precipitation is used to treat various aqueous hazardous and industrial wastes for the removal of substances such as soluble heavy metal, dissolved organics and colloidal particles suspended in liquid. Colloidal particles are very fine dispersed solids suspended in liquid streams and are stabilized by negative surface charges. The negative charges formed make the colloids to repel from one another, avoiding the particles from forming larger masses, called as flocs. Since the colloidal particles suspended in the liquid and cannot settle down on their own, the application of coagulation and flocculation treatment is necessary. A combination of coagulation, flocculation, precipitation and sedimentation processes are frequently used to maximize suspended solids removal. Typically, the metal precipitation in the solution is in the hydroxide form.

Over the past few years, several authors have investigated the role of few chemicals such as hydrated lime or Calcium Hydroxide ($Ca(OH)_2$), alum, magnesium hydroxide ($Mg(OH)_2$), Polyaluminum Chloride (PaCl), Magnesium Ammonium Phosphate (MAP) and sodium hydroxide or caustic soda (NaOH) as potential chemicals for the clarifications of most wastewater and leachate (Cheung et al., 1997; Shabiimam & Dikshit, 2011; Zainol et al., 2011; Imran et al., 2012). Among the chemicals available, hydrated lime ($Ca(OH)_2$) and struvite (magnesium ammonium phosphate (MAP)) are generally employed.

Sani et al. (2014) evaluated the removal of NH_3-N from raw SWL leachate using lime precipitation at different lime concentration (2, 4, 6, 8 and 10 gL^{-1}) by jar test experiment. The result indicated that NH_3-N removal increases up to 54% with 4 gL^{-1} of optimum dosage at pH 12.39. However, further addition of lime at 6 gL^{-1} has no significant effect on NH_3-N removal.

5.2. Ammonia/Air Stripping

Air stripping or ammonium stripping is a mass transfer operation involving the transfer of dissolved substances (pollutants) in water from liquid phase to gas phase especially volatile organic compounds (VOCs), where air and water interact in a column which is specifically designed to maximize the contact surface area between the liquid (water/wastewater) and gas (air) (Beranek, 2001; Huang and Shang, 2006; El-Behlil and Adma, 2012; Abdullahi et al., 2013; Abdullahi, 2014). It is widely used for removing ammonia and other VOCs. VOCs are organic compounds which have a boiling point less than 100 °C and/or vapour pressure greater than 1 mm Hg at 25 °C (Metcalf and Eddy, 1991). Generally, leachate contains high concentration of ammonium-nitrogen (NH_4^+-N), which may cause toxicity of leachate, and stripping can be successful for removing this pollutant (Marttinen et al., 2002). The process of air stripping or ammonia stripping involved the removal of ammonia (NH_3-N or NH_4^+-N) from liquid wastewater by volatilization of gaseous ammonia (Vigneswaran, 2009). Additionally, it is often easier and less expensive to remove ammonia–nitrogen (NH_3–N) in the form of ammonia gas (NH_3) rather than to convert it to nitrate-nitrogen before discharged to the environment.

Ammonia reacts with water to form ammonium ion (NH_4^+) and hydroxide ion (OH^-). It is often referred to as liquid ammonia, aqueous ammonia or ionized ammonia. Ammonia in gaseous form (NH_3) is toxic to aquatic organisms. However, ammonium ion (NH_4^+) is non-toxic. An equilibrium exists between the toxic NH_3 and non-toxic NH_4^+, as shown in Eq. (1) is reversible:

$$NH_3\,(g) + H_2O\,(l) \leftrightharpoons NH_3 \cdot H_2O\,(aq) \leftrightharpoons NH_4^+\,(aq) + OH^-\,(aq) \qquad (1)$$

Based on Eq. (1), ammonium ion (NH_4^+) and free ammonia (NH_3) concentrations depend on pH. At lower pH, ammonium and hydrogen ions are dominant species. As the pH increases, the equation shifts to the left and consequently a higher elevation of the gaseous fraction. With a pKa of 9.25 of ammonium/ammonia equilibrium, ammonium nitrogen can be easily transferred from a liquid to dissolve ammonia gas. As an alkaline agent such as lime added into the leachate, the pH of leachate will increase, hence more NH_4^+ will be converted to NH_3.

Factors that influence the dynamic equilibrium between NH_3 and NH_4^+ are pH and water temperature. At pH 6, the ratio of NH_3 to NH_4^+ is 1: 3000. This ratio decreases to 1:30 when the pH increases to 8 (alkaline). In addition, as the temperature increases, the ammonia toxicity also increases (Wurts, 2003). Usually, ammonium ion (NH_4^+) is in the form of NH_3 gas at pH 11 or 12 (Bonmati and Flotats, 2003; Kurniawan et al., 2006). Therefore, air stripping process requires a pH adjustment of the leachate, which is by increasing the pH within the range of 10.5 to 12 to convert the ammonia in leachate into gaseous form (Diamadopoulos, 1994; Ozturk et al., 2003; Calli et al., 2005). At pH 7 or below, the ammonia is in soluble form, ammonium (NH_4^+). As the pH of leachate is higher, which is around 11 or 12, air stripping is much more effective especially for removal of ammonia (NH_3) (Bonmatí and Flotats, 2003). Generally, calcium hydroxide (Diamadopoulos, 1994; Ozturk et al., 2003; Calli et al., 2005; Cotman and Gotvajn, 2010; Guo et al., 2010) or sodium hydroxide (NaOH) (Marttinen et al., 2002; Silva et al., 2004) are used as the added alkali to increase the pH.

There are two main types of air stripping units which are countercurrent flow unit and diffused aerator unit. In a countercurrent flow unit, air is blown from the bottom while the leachate is applied over a stripping tower (cylindrical column) that is packed with packing materials. The plastic packing materials, which act as a media in the tower increases the surface area of the leachate that is exposed to air. During the air stripping process, as the leachate moves down the tower, the large quantities of air flowing against the leachate flow will strip away the VOCs and ammonia gas (NH_3) allowing them to escape from the liquid state into the air stream in the form of gas. The purified water or wastewater from the treatment is collected at the bottom of the column while the stripping gas (VOCs) leaves the column at the top (Mead and Leibbert, 1998; Beranek, 2001; Richardson et al., 2002; Mourad et al., 2012). The example of countercurrent air stripper unit includes packed bed and sieve tray air strippers.

Packed column air strippers operate effectively over a narrow range of water flows of 0.8 to 1.8 m^3/min. This means less air used for the same water flow rate by packed column air stripper compared to sieve tray air stripper. In addition, packed column air stripper able to operate in wide range of flow rates (1.5 to 76 m^3/min), which means that when water flow rate to the column decreases, air flow rate will also decrease. As air flow rate decrease, less energy is required, hence reducing the cost for air stripping treatment. Packed column air strippers also exhibit lower pressure drop characteristics than the pressure drop through sieve tray air strippers. This is an advantage of packed column air strippers as smaller blower and motor can be used, which consequently reduced the cost of electricity. To conclude, packed column air strippers are more favoured as they are easier to construct and are also more economical (Abdullahi, 2014).

6. Case Study: Treatment of Scheduled Waste Landfill Leachate by Lab-Scale Ammonia Stripping

6.1. Sample Collection and Characterization

Leachate samples were collected from a Scheduled Waste Landfill (SWL) facility in Peninsular Malaysia. The scheduled waste landfill started its operation in the year 2000. The major constituents of the scheduled waste are the incinerated ashes of industrial and hazardous waste from factories, hospitals and others. The facility is equipped with a leachate collection pond. The leachate samples were collected in May 2014. The collection and preservation of samples was undertaken in accordance with the Standard Methods for the Examination of Water and Wastewater (APHA, 2005). The leachate samples were collected in a clean 25 L containers, and immediately transported to the laboratory to be properly stored in a refrigerator at 4°C. The characteristics of SWL as shown in Table 12.

The average pH of SWL (10.0) was very alkaline. The BOD_5 /COD ratio describes the degree of biodegradation and gives information about the landfill age. Generally, low BOD_5/COD ratio shows the high concentration of non-biodegradable organic compounds, which makes the leachate difficult to be biologically degraded. For the SWL, the BOD_5/COD ratio was 0.50. Schedule waste landfill was considered as young (<5 years) since the BOD_5/COD ratio value was more than 0.3.

Table 12. Characteristics of raw leachate from Scheduled Waste Landfill

Parameter	Units	SWL leachate		DOE Standard*
		Range	Average	
Temperature	°C	20.3 - 20.8	20.5	40
pH Value	-	9.7-10.2	10.0	5.5-9.0
Biochemical Oxygen Demand (BOD_5)	mgL^{-1}	1610 - 2670	2053	50
Chemical Oxygen Demand (COD)	mgL^{-1}	3430 - 4610	4127	100
Total Suspended Solids (TSS)	mgL^{-1}	0 - 870	313	100
Copper	mgL^{-1}	0.18 – 1.04	0.55	1.0
Arsenic	mgL^{-1}	9.24 - 16.9	13.01	0.10
Boron	mgL^{-1}	27.2 – 31.0	29.30	4.0
Phenols (Total)	mgL^{-1}	0.730 – 8.08	3.26	1.0
Ammoniacal-Nitrogen (NH_3-N)	mgL^{-1}	1790 - 2570	2267	20
Colour (pH Natural)	ADMI	52-1590	1004	100
Colour (pH adjusted to 7.6)	ADMI	50-1550	913	100

* DOE Standard: Environmental Quality (Control of Pollution from Solid Waste Transfer Station and Landfill) Regulations 2009, Second Schedule (Standard B).

6.2. Pre-Treatment by Chemical Precipitation

A recent study by Hanira et al. (2017a) investigated the potential use of hydrated lime ($Ca(OH)_2$) and caustic soda (NaOH) as chemicals for increasing the pH prior ammonia removal by an ammonia-stripping column. Coagulation studies were performed using a jar test apparatus, with six impellers equipped with 2.5 cm×7.5 cm rectangular blades and six beakers (VELP-Scientifica, Model: JLT6, Italy). The time and speed for rapid and slow mixing were set using an automatic controller.

Different dosages and pH were investigated for $Ca(OH)_2$ and NaOH in removing chemical oxygen demand (COD), colour and ammonia-nitrogen

(NH_3-N). Each beaker containing leachate was applied with 0 gL^{-1} (as a control) to 12 gL^{-1} of $Ca(OH)_2$, respectively. The operating parameters were adopted as a rapid mixing speed of 80 rpm for a period of 2 minutes, followed by a slow mixing speed of 30 rpm for 30 minutes with an additional 120 minutes of settling time. The pH, ammonia-nitrogen (NH_3-N), chemical oxygen demand (COD), and colour were measured. The parameter measurements were conducted based on the Standard Methods for the Examination of Water and Wastewater (APHA, 2005). The pH was measured using a portable instrument (Hanna Instruments, Model: 991003, Romania). The ammonia-nitrogen (NH_3-N) in the sample was measured using Ammonia Ion Selective Electrode (HACH, Intellical). COD was determined using the colorimetric method (5220-D). Colour was measured as true colour based on APHA Platinum-Cobalt Standard Method using a DR 890 HACH colorimeter. Method Number 8025 reports the colour in platinum–cobalt (Pt–Co). The effect of filtration on colour removal was corrected by means of a control sample.

The removal efficiency for the parameters measured was calculated using the following formula:

$$\text{Removal (\%)} = 100 \text{ x } [(C_i - C_f)/C_i] \quad (2)$$

where C_i and C_f are the initial and final concentrations of the parameters measured in the leachate sample in mgL^{-1}, respectively (Hanira et al., 2015).

As depicted in Figure 1, the concentration of $Ca(OH)_2$ and NaOH used were ranging from 0 gL^{-1} to 12 gL^{-1}. The maximum removal of NH_3-N was observed at Ca $(OH)_2$ concentrations of 5 gL^{-1} and NaOH at 8 gL^{-1} with 61% and 42%, respectively. The study showed that less amount of $Ca(OH)_2$ was needed (i.e., 5 gL^{-1}) to obtain the optimum NH_3-N removal efficiency compared to NaOH (i.e., 8 gL^{-1}). The pH value recorded at the optimum dosage of 5 gL^{-1} Ca $(OH)_2$ and 8 gL^{-1} NaOH were pH 12.17 and 12.83, respectively. Further addition of coagulants than the optimum dosage showed a slight decrease in the NH_3-N removal.

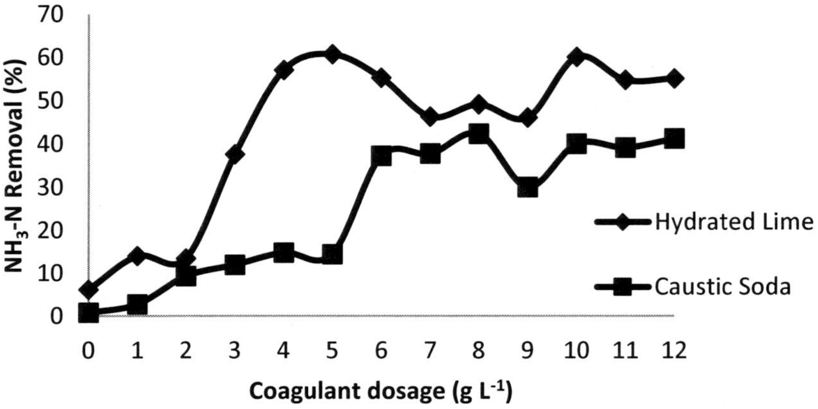

Figure 1. Removal of ammonia-nitrogen at various dosages of hydrated lime, Ca $(OH)_2$ and caustic soda (NaOH).

Results showed that $Ca(OH)_2$ is the most suitable chemical for the pretreatment of SWL leachate. Previous research had also demonstrated the usage of $Ca(OH)_2$ as an added alkali aids in removal of suspended solids, ammonia-nitrogen (NH_3-N), colour, heavy metals and other substances. (Cheung et al., 1997; Ozturk et al., 2003; Mirbagheri and Hosseini; 2004; Malakootian and Fatehizadeh; 2010; Shabiimam and Dikshit, 2011). In Calli et al. (2005) study, lime was preferred over NaOH due to excessive foaming when NaOH was used during pH adjustment. In fact, $Ca(OH)_2$ required less dosage in their study. Thus, $Ca(OH)_2$ could be used as an effective pre-treatment step prior to NH_3-N removal in a Scheduled Waste Landfill (SWL) leachate treatment plant.

6.3. Treatment by Air Stripping Column

Recently, Hanira et al. (2017b) studied the effect of $Ca(OH)_2$ dosage, pH and air flow rate on the NH_3-N removal from a scheduled waste landfill (SWL) leachate using a lab-scale air stripping column, by employing a central composite design (CCD) of response surface methodology (RSM) to optimize the parameters affecting NH_3-N removal. Figure 2 depicted the experimental set-up of the lab-scale air stripping column made of PVC of 260 mmH x 100 mmD with a total capacity of 2.0 L and a working volume of 1.0 L. Initially, pH of leachate was adjusted to 10-12 by 0.1 M sodium hydroxide solutions (Wang et al., 2007). The SWL leachate then was pre-treated with hydrated lime ($Ca(OH)_2$) at varying dosage ranging from 0-10 gL^{-1} in each separate experiment to enhance NH_3-N removal in SWL leachate. Air was supplied through an air diffuser at the bottom of the column at various airflow rates ranging from 1.5 - 2.5 L min^{-1}. The leachate sample was aerated for 24 hours and the concentrations of NH_3-N were measured on an hourly basis using Ammonia Ion Selective Electrode (HACH, Intellical as was shown in Eq. (2).

For this chapter, the Design Expert Software (version 7.0) was adopted for the statistical design of experiments and data analysis. The optimization of $Ca(OH)_2$ dosage, initial pH and air flow rates on the removal of NH_3-N was performed using Central composite design (CCD) of the RSM. The operating parameters used in the study were as follows: lime dosage 0-10 gL^{-1}, initial pH of 10-12, and air flow rate of 1.5- 2.5 L min^{-1}. The results for quadratic regression model of NH_3-N removal efficiency in terms of code factor was obtained as Eq. (3) below:

$$Y_{NH3\text{-}N} = 91.21 + 1.31\,X_1 + 6.06\,X_2 + 5.40\,X_3 + 0.85\,X_1X_2 + 1.45\,X_1X_3 + 1.03\,X_2X_3 + 0.85\,X_1^2 - 12.20\,X_2^2 - 2.80\,X_3^2 \quad (3)$$

Based on the regression equation, the optimal conditions for NH_3-N removal obtained were: $Ca(OH)_2$ dosage= 9.8 g L^{-1}, initial pH= 11 and air flow rate= 2.25 L min^{-1}. With NH_3-N removal as the response, the response surfaces of the quadratic model were shown in Figure 3 to Figure 5. As depicted in Figure 3, $Ca(OH)_2$ dosage and initial pH have strong effects on NH_3-N removal efficiency, where the optimum value identified was $Ca(OH)_2$ dosage of 5 to 10 g L^{-1} and an initial pH of 11-11.5. An increase in pH of leachate increases the fraction of gaseous NH_3, leading to the release of NH_3 to the air, and consequently a reduction in the NH_3-N concentration in liquid leachate.

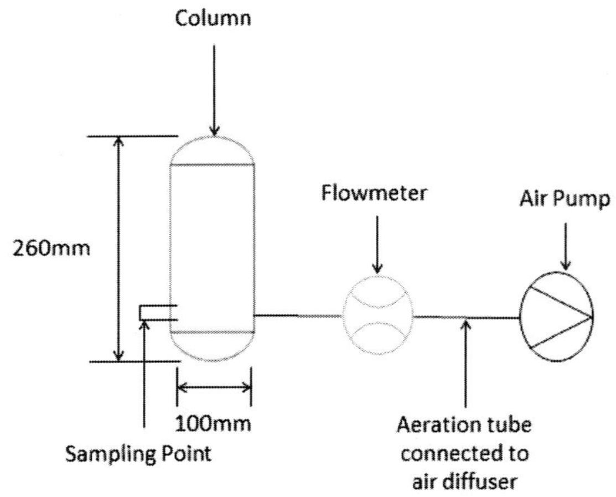

Figure 2. The setup for air stripping column.

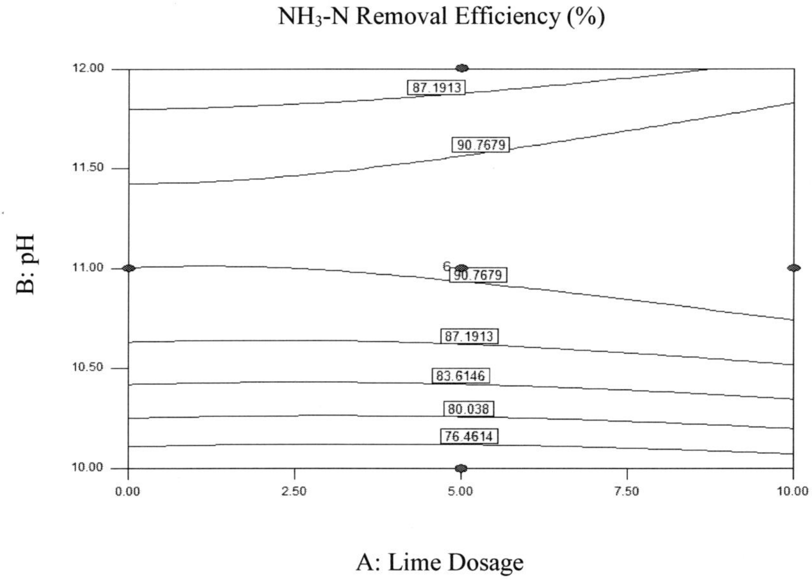

Figure 3. Design-expert plot; contour plot for NH_3-N removal at different lime dosage and initial pH.

Figure 4 and Figure 5 represents the contour plots for NH_3-N removal at different $Ca(OH)_2$ dosage and an air flow rate, and different initial pH and air flow rate, respectively. $Ca(OH)_2$ dosage varying from 5 to 10 g L^{-1}, initial pH ranging from 11 to 11.5 and air flow rate ranging from 2 to 2.5 L min^{-1} were found to be the most optimum for this study. The findings from this study was similar to those reported earlier, which indicates higher air flow rate enhanced the NH_3-N removal performances for landfill leachate (Cheung et al., 1997; Lei et al., 2007; Yilmaz et al., 2010).

The second order polynomial model was used to confirm the validity of the statistical experimental strategies at optimum conditions for $Ca(OH)_2$ dosage, initial pH and air flow rate. The results from RSM proposed the optimal conditions for maximum removal of NH_3-N were at lime dosage of 9.8 g L^{-1}, initial pH 11 and air flow rate of 2.25 L min^{-1} as in Table 13. It was estimated that with these optimum conditions, the maximum value for NH_3-N removal via air stripping was 96.5%. Verification experiments was carried out thrice and the results obtained was 96.3% of NH_3-N removal (influent= 1347 mg L^{-1}, effluent= 49.5 mg L^{-1}), with final pH of 10.21 after 24 hours.

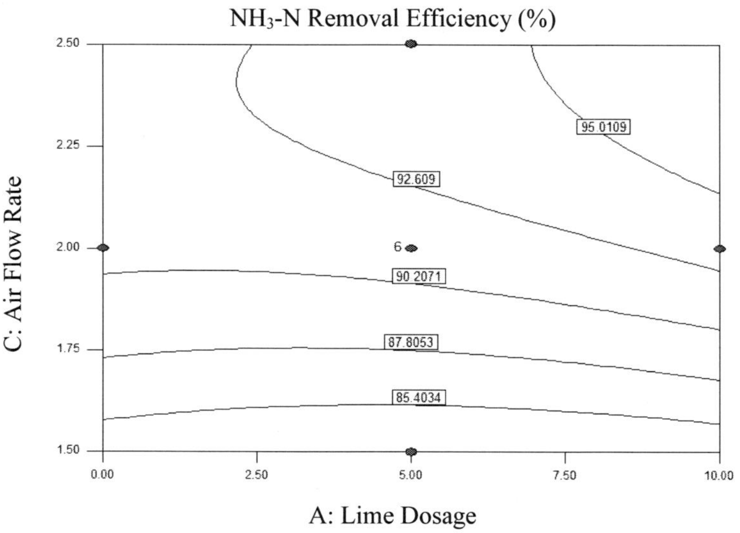

Figure 4. Design-expert plot; contour plot for NH_3-N removal at different lime dosage and air flow rate.

Previous researchers observed that an air flow rate of 5 L min^{-1} with 10 g L^{-1} $Ca(OH)_2$ can adequately remove up to 90% of NH_3-N after 24 hours of treatment (Cheung et al., 1997). In Ozturk et al. (2003) study, with an aeration rate of 7.6 L min^{-1} and 8 g L^{-1} of $Ca(OH)_2$, pH of young municipal landfill leachates increases until pH 12, and 85% of ammonia removal efficiency was obtained after 17 hours (Ozturk et al., 2003). Compared to this work, at an initial pH of 11, 9.8 g L^{-1} of $Ca(OH)_2$ dosage and aeration treatment at a rate of 2.25 L min^{-1}, 96.4% of NH_3-N were successfully removed from scheduled waste leachate within 24 hours. This shows that air stripping process can be used efficiently for removal of NH_3-N in scheduled waste landfill leachate. Based on the results, RSM approach could be used for the optimization of the operational conditions.

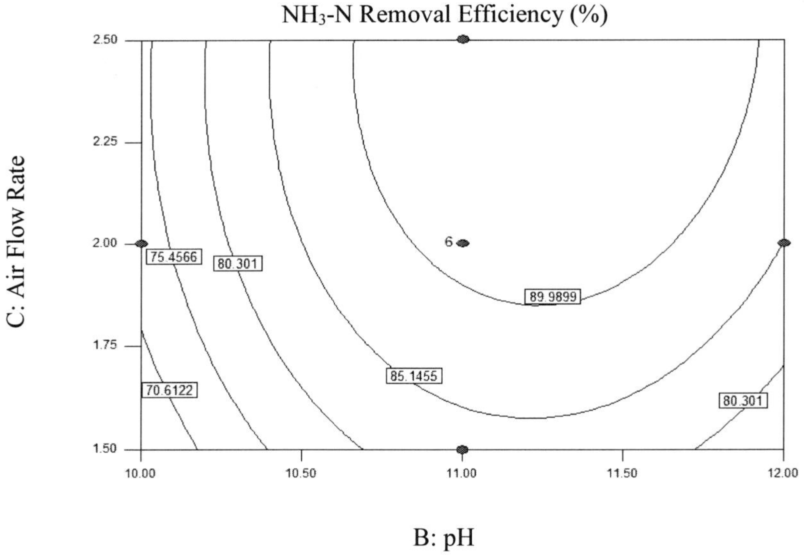

Figure 5. Design-expert plot; contour plot for NH_3-N removal at different initial pH and air flow rate.

Table 13. The optimum parameters and NH_3-N removal obtained by RSM

Lime Dosage (g L^{-1})	Final pH	Air Flow Rate	NH_3-N removal (%)	
			Actual	Predicted
9.8	10.21	2.25	96.3	96.5

CONCLUSION

This chapter gives a comprehensive study on scheduled waste landfill leachates in Malaysia. The fundamentals and techniques of treating SWL in Malaysia were also presented. This study revealed that a combination of either biological, chemical and physical treatment method is most efficient in treating SWL. Chemical precipitation (54%) and air-stripping methods (90-97%) have also been found suitable in removing the high toxic compounds from SWL. For increased efficiency of SWL treatment, optimum parameters must be applied. The response surface methodology (RSM) has been demonstrated as an adequate method for the optimization of SWL treatment parameters. Overall, the information and data presented in this chapter are meant to introduce and familiarize the reader with the management of schedule waste landfill leachate. Along with the case study to provide design criteria and operational ideas based on laboratory studies, so that the readers are able to consider and compare the pro and cons of the treatment available.

REFERENCES

Abbas, A. A., Jingsong, G., Ping, L. Z., Pan, Y. Y., Al-Rekabi, W. S. (2009). Review on Landfill Leachate Treatments. *Journal of Applied Sciences Research,* 5, 534–545.

Abdullahi, M. E., Hassan, M. A. A., Noor, Z. Z., Raja Ibrahim, R. K. (2013). World Simulation of the Effect of Process Variables on Packed Column Air Stripper Performance. *World Applied Sciences Journal,* 25 (7), 1100-1106.

Abdullahi, M. E., Hassan, M. A. A., Noor, Z. Z., Raja Ibrahim, R. K. (2014). Application of a Packed Column Air Stripper in the Removal of Volatile Organic Compounds from Wastewater. *Reviews in Chemical Engineering,* 30(5), 431–451.

Aja, O. C., Al-Kayiem, H. H., Zewge, M. G., Joo, M. S. (2016). Overview of Hazardous Waste Management Status in Malaysia in *Management of Hazardous Wastes*. Chapter 5 (Open Access). DOI: 10.5772/63682.

Al-Hamadani, Y. A. J., Yusoff, M. S., Umar, M., Bashir, M. J. K., and Adlan, M. N. 2011. Application of psyllium husk as coagulant and coagulant aid in semi-aerobic landfill leachate treatment. *Journal of Hazardous Materials.* 190: 582-587.

Aluko, O. O., and Sridhar, M. K. (2013). Evaluation of Leachate Treatment by Trickling Filter and Sequencing Batch Reactor Processes in Ibadan, Nigeria. *Waste Management & Research*, 31(7), 700-5. doi: 10.1177/0734242X13485867.

Alvarez-Vazquez, H., Jefferson, B., Simon, J. J. (2004). Membrane Bioreactors vs Conventional Biological Treatment of Landfill Leachate: A Brief Review. *Journal of Chemical Technology and Biotechnology,* 79 (10), 1043 - 1049.

APHA. 2005. *Standard methods for the examination of water and wastewater* (21st ed.). Washington, DC: American Public Health Association.

Aziz, H. A., Yussoff, M. S., Adlan, M. N., Adnan, N. H., Alias, S. (2004). Physico-chemical Removal of Iron from Semi-aerobic Leachate by Limestone Filter. *Waste Management,* 24 (4), 353-358.

Aziz, S. Q., Aziz, H. A., Bashir, M. J. K., and Mojiri, A. (2015). Assessment of Various Tropical Municipal Landfill Leachate Characteristics and Treatment Opportunities, *Global NEST Journal, 17,* 1-13.

Azmi, N., Bashir, M. J. K., Sethupathi, S., and Ng, C. A., (2014). Anaerobic Stabilized Landfill Leachate Treatment using Chemically Activated Sugarcane Bagasse Activated Carbon: Kinetic and Equilibrium Study. *Desalination and Water Treatment*, 57 (9), 3916-3927.

Beranek, D. (2001). *Engineering and design air stripping: Design Guide No. 1110-13* Washington, D.C.: Department of Army, U.S. Army corps of engineers. (pp: AI-1 to D18).

Berrueta, J., and Castrillón, L. (1992). Anaerobic Treatment of Leachates in UASB Reactors. *Journal of Chemical Technology and Biotechnology,* 54, 33–37.

Białowiec, A. (2011). Hazardous Emissions from Municipal Solid Waste Landfills. In *Contemporary Problems of Management and Environmental Protection,* No. 9. Some Aspects of Environmental Impact of Waste Dumps (pp. 7-28).

Blauvelt, A. (2009). *Removal of Ammonia from Landfill Leachate.* (Bachelor Science Thesis). Worcester Polytechnic Institute.

Bonmatí, A., and Flotats, X. (2003). Air Stripping of Ammonia from Pig Slurry: Characterisation and Feasibility as a Pre- or Post-Treatment to Mesophilic Anaerobic Digestion. *Waste Management,* 23, 261-272.

Cai, T., Park, S. Y., and Li, Y. (2013). Nutrient Recovery from Wastewater Streams by Microalgae: Status and Prospects. *Renewable and Sustainable Energy Reviews,* 19, 360-369.

Calli, B., Mertoglu, B., Inane, B. (2005). Landfill Leachate Management in Istanbul: Applications and Alternatives. *Chemosphere,* 59, 819 – 829.

Cheung, K. C., Chu, L. M., Wong, M. H. (1997). Ammonia Stripping as a Pretreatment for Landfill Leachate. *Water, Air and Soil Pollution,* 94, 209-221.

Choo, K. H., and Lee, C. H. (1996). Membrane Fouling Mechanisms in the Membrane-coupled Anaerobic Reactor. *Water Research,* 30 (8), 1771-1780.

Clément, B., and Merlin, G. (1995). The Contribution of Ammonia and Alkalinity to Landfill Leachate Toxicity to Duckweed. *Science of the Total Environment,* 170 (1-2), 71-79.

Cotman, M. I., and Gotvajn, A. Z. (2010). Comparison of Different Physico-chemical Methods for the Removal of Toxicants from Landfill Leachate. *Journal of Hazardous Materials*, 178 (1-3), 298-305.

Diamadopoulos, E. (1994). Characterization and Treatment of Recirculation-Stabilized Leachate. *Water Research,* 28 (12), 2439-2445.

Di Iaconi, C., Ramadori, R., & Lopez, A. (2006). Combined biological and chemical degradation for treating a mature municipal landfill leachate.

Biochemical Engineering Journal, 31(2), 118–124. https://doi.org/10.1016/j.bej.2006.06.002.

DOE (2012) *Guidance for the industry on the notification and registration scheme of Environmentally Hazardous Substances (EHS) in Malaysia.* Department of Environment, Ed.

DOE (2015) *EHS Reference List* [Online]. Available: https://www.e-ehs.doe.gov.my.my/public/references/index.

El-Behlil, M. A., and Adma, S. A. (2012). Volatile Organic Chemicals Removal from Contaminated Water using Low Profile Sieve Tray Air Stripping Towers. *International Water Technology Journal,* 2 (2), 164 – 176.

El-Fadel, M., Bou-Zeid, E., Chahine, W., & Alayli, B. (2002). Temporal variation of leachate quality from pre-sorted and baled municipal solid waste with high organic and moisture content. *Waste Management, 22*(3), 269–282. https://doi.org/10.1016/S0956-053X(01)00040-X.

El-Gohary, F. A., Khater, M., Gamal, M. K. (2013). Pretreatment of Landfill Leachate by Ammonia Stripping. *Journal of Applied Sciences Research,* 9 (6), 3905-3913.

Fettig, J., Stapel, H., Steinert, C., and Geiger, M. (1996). Treatment of Landfill Leachate by Preozonation and Adsorption in Active Carbon Columns. *Water Science and Technology,* 34 (9), 33-40.

Ferraz, F. M., Povinelli, J., Vieira, E. M., (2013). Ammonia Removal from Landfill Leachate by Air Stripping and Absorption. *Environmental Technology,* 34, 2317-2326.

Gurijala, K. R., and Suflita, J. M. (1993). Environmental Factors Influencing Methanogenesis from Refuse in Landfill Samples. *Environmental Science & Technology,* 27 (6), 1176–1181.

Hamzah, U., Jeeva, M., and Mohd Ali, N. A. (2014). Electrical Resistivity Techniques and Chemical Analysis in the Study of Leachate Migration at Sungai SEDU Landfill. *Asian Journal of Applied Sciences*, 7 (7), 518-535. DOI: 10.3923/ajaps.2014.518.535.

Hanira, N. M. L., Hasfalina C. M., Sani, A., and Rashid, M. (2017a). Pre-treatment Ammonia Removal of Scheduled Waste Leachate with Hydrated Lime and Caustic Soda. *Jurnal Teknologi*, 79 (1), 107–118.

Hanira, N. M. L., Hasfalina C. M., Rashid, M., Luqman C. A., and Abdullah, A. M. (2017b). Removal of Ammonia-Nitrogen (NH_3-N) from Scheduled Waste Leachate by Air Stripping, *Desalination and Water Treatment*, 68, 330-337.

Hanira, N. M. L., Hasfalina, C. M., Sani, A., Rashid, M. (2015). Comparison of Lime Powder and Caustic Soda as a Pre-treatment for Ammonia-nitrogen Removal from a Scheduled Waste Leachate. *AIP Conference Proceeding,* 1660, 090010-1–090010-8. doi: 10.1063/ 1.4915854.

Hoilijoki, T. H., Kettunen, R. H., Rintala, J. A. (2000). Nitrification of Anaerobically Pretreated Municipal Landfill Leachate at Low Temperature. *Water Research,* 34 (5), 1435-1446.

Horan, N. J., Gohar, H., Hill, B. (1997). Application of a Granular Activated Carbon-Biological Fluidized-Bed for the Treatment of Landfill Leachate containing High Concentrations of Ammonia. *Water Science and Technology,* 36, 369-375.

Hossini, H., Rezaee, A., Ayati, B., Mahvi, A. H. (2015). Optimizing Ammonia Volatilization by Air Stripping from Aquatic Solutions using Response Surface Methodology (RSM). *Desalination and Water Treatment,* 57 (25), 11765-11772.

Huang, J., and Shang, C. (2006). Air Stripping. In Wang, L. K., Hung, Y. and Shammas, K. N. (Eds.) *Hand book of Environmental Engineering: Advanced Physicochemical processes* (Vol. 4, pp. 47-78). New Jersey: Humana Press Incorporated.

Imran, Q., Hanif, M. A., Riaz, M. S., Noureen, S., Ansari, T. M., Bhatti, H. N. (2012). Coagulation/Flocculation of Tannery Wastewater Using Immobilized Chemical Coagulants. *Journal of Applied Research and Technology,* 10 (2), 79-86.

Jamin, N. C., and Mahmood, N. Z. (2015). Scheduled Waste Management in Malaysia: An Overview. *Advanced Materials Research,* 1113, 841-846.

Jokela, J. P. Y., Kettunen, R. H., Sormunen, K. M., Rintala, J. A. (2002). Biological Nitrogen Removal from Municipal Landfill Leachate: Low-

Cost Nitrification in Biofilters and Laboratory Scale in-situ Denitrification. *Water Research,* 36 (16), 4079- 4087.

Kennedy, E. J., and Lentz, E. M. (2000). Treatment of Landfill Leachate using Sequencing Batch and Continuous Flow Upflow Anaerobic Sludge Blanket (UASB) Reactors. *Water Research,* 34 (14), 3640–3656.

Kettunen, A., Kaitala, V., Alm, J., Silvola, J., NykaÈnen, H., Martikainen, P. J. (1996). Cross-correlation Analysis of the Dynamics of Methane Emissions from a Boreal Peatland. *Global Biogeochemical Cycle,* 10, 457-471.

Kjeldsen, P., Barlaz, M. A., Rooker, A. P., Baun, A., Ledin, A., Christensen. T. H. (2002). Present and Long-term Composition of MSW Landfill Leachate: A Review. *Criteria Reviews in Environmental Science and Technology,* 32, 297-336.

Kurniawan, T. A., Lo, W. H., Chan, G. Y. S. (2006). Physico-chemical Treatments for Removal of Recalcitrant Contaminants from Landfill Leachate. *Journal of Hazardous Materials*, B129, 80 – 100.

Kurniawan, T. A. (2011). *Treatment of Landfill Leachate.* Lap Lambert Academic publishing GmbH and Co. KG.

Kutty, S. R. M., Isa, M. H., Leong, L. C. (2011). Removal of Ammonia-Nitrogen (NH_3-N) and Nitrate (NO_3^-) by Modified Conventional Activated-Sludge System to Meet New D.O.E Regulations. *International Conference on Environment and Industrial Innovation IPCBEE.* 12, 103-107.

Lei, X., Sugiura, N., Feng, C., and Maekawa, T. (2007). Pretreatment of Anaerobic Digestion Effluent with Ammonia Stripping and Biogas Purification. *Journal of Hazardous Materials*, 145, 391-397.

Lema, J. M., Mendez, R., Blazquez, R. (1988). Characteristics of Landfill Leachates and Alternatives for Their Treatment: A review. *Water, Air and Soil Pollution,* 40, 223-250.

Li, X. Z., Zhao, Q. L., Hao, X. D. (1999). Ammonium Removal from Landfill Leachate by Chemical Precipitation. *Waste Management,* 19, 409-415.

Lin, S. H., and Chang, C. H. (2000). Treatment of Landfill Leachate by Combined Electro-Fenton Oxidation and Sequencing Batch Reactor Method. *Water Research,* 34, 4243-4249.

Lin, S. H., and Wu, C. L. (1996). Removal of Nitrogenous Compounds from Aqueous Solution by Ozonation and Ion Exchange. *Water Research,* 30, 1851-1857.

Liu, S. (2013). *Landfill Leachate Treatment Methods and Evaluation of Hedeskoga and Måsalycke Landfills.* (Master Thesis). Lund University.

Loukidou, M. X., and Zouboulis, A. I. (2001). Comparison of Two Biological Treatment Process Using Attached Growth Biomass for Sanitary Landfill Leachate Treatment. *Environmental Pollution,* 111 (2), 273-281.

Lopez, A., Pagano, M., Volpe, A., Di Pinto, A. (2004). Fenton's Pre-treatment of Mature Landfill Leachate. *Chemosphere,* 54, 1005-1010.

Lucas, J. M. (1994). How to Achieve a Robust Process using Response Surface Methodology. *Journal of Quality Technology,* 24 (6), 248 – 260.

Malakootian, M., and Fatehizadeh, A. (2010). Color Removal from Water by Coagulation/Caustic Soda and Lime. *Iranian Journal of Environmental Health Science & Engineering,* 7, 267-272.

Malaysia Environmental Quality Report 2015. (2016). Department of Environment, Ministry of Natural Resources and Environment. Omr Press Sdn. Bhd. Selangor, Malaysia. 138 – 149.

Mårtensson, A. M., Aulin, C., Wahlberg, O., Ågren, S. (1999). Effect of Humic Substances on the Mobility of Toxic Metals in a Mature Landfill. *Waste Management and Research,* 17(4), 296-304.

Marttinen, S. K., Kettunen, R. H., Sormunen, K. M., Soimasuo, R. M., Rintala, J. A. (2002). Screening of Physical-chemical Methods for Removal of Organic Material, Nitrogen and Toxicity from Low Strength Leachates. *Chemosphere,* 46, 851-858.

Maynard, H. E., Ouki, S. K., William, S. C. (1999). Tertiary Lagoons: A Review of Removal Mechanisms and Performance. *Water Research,* 33, 1-13.

Mead, E., and Leibbert, J. (1998). A Comparison of Packed Column and Low Profile Sieve Tray Strippers. *Proceedings of the 1998 Conference on Hazardous Water Research*. Kansas, USA. 328 – 334.

Metcalf and Eddy. (1991). *Wastewater Engineering, Treatment, Disposal and Reuse*. 3rd Ed. pp. 359–756. New York: McGraw-Hill.

Mirbagheri, S. A., and Hosseini, S. N. (2004). Pilot Plant Investigation on Petrochemical Wastewater Treatment for the Removal of Copper and Chromium with the Objective of Reuse. *Desalination,* 171, 85-93.

Moraes, P. B., and Bertazzoli, R. (2005). Electrodegradation of Landfill Leachate in a Flow Electrochemical Reactor. *Chemosphere,* 58, 41-46. doi: 10.1016/j.chemosphere.2004.09.026.

Mourad, K., Berndtsson, R., Abu-Elsha'r, W., Qudah, M. A. (2012). Modelling Tool for Air Stripping and Carbon Adsorbers to Remove Trace Organic Contaminants. *International Journal of Thermal and Environmental Engineering,* 4 (1), 99 – 106.

Murunga, S. I., Mbuge, D. O., Gitau, A. N., Mutwiwa, U. N., and Wekesa, I. N. (2016). Isolation and Characterization of Methanogenic Bacteria from Brewery Wastewater in Kenya. *African Journal of Biotechnology*, 15 (47), 2687-2697. DOI: 10.5897/AJB2016.15551

Olsson, M. P., and Lo, K. V. (1997). Sequencing Batch Reactors for the Treatment of Egg Processing Wastewater. *Canadian Agricultural Engineering*, 39 (3), 195–202.

Ozturk, I., Altinbas, M., Koyuncu, I., Arikan, O., Gomec-Yangin, C. (2003). Advanced Physico-chemical Treatment Experiences on Young Municipal Landfill Leachates. *Waste Management,* 23(5), 441-446.

Qasim, S. (1994). *Wastewater Treatment Plants: Planning, Design and Operation*. Technomic Publishing Co., Lancaster, Pennsylvania.

Renou, S., Poulain, S., Givaudan, J. G., Moulin, P. (2008). Treatment Process Adapted To Stabilized Leachates: Lime Precipitation-Pre-filteration-Reverse Osmosis. *Journal of Membrane Science,* 313, 9 – 22.

Renou, S., Givaudan, J. G., Poulain, S., Dirassouyan, F., & Moulin, P. (2008). Landfill leachate treatment: Review and opportunity. *Journal of Hazardous Materials*, *150*(3), 468–493. https://doi.org/10.1016/j.jhazmat.2007.09.077

Richardson, J. F., Harker, J. H., Backhurst, J. R. (2002). *Particle Technology and Separation Processes*. London: Butterworth and Heinamann.

Salleh, N. F. D. M and Hamid, K. H. K. (2013). Leachate Characterization from a Closed Landfill in Air Hitam, Puchong, Malaysia. *The Malaysian Journal of Analytical Sciences*, 17 (1), 24-29.

Sani, A., Rashid, M., Nurul Hanira, M. L., Hasfalina, C. M. (2014). The Influence of pH on the Removal of Ammonia from a Scheduled Waste Landfill Leachate. *Jurnal Teknologi,* 68 (5), 25-28.

Shabiimam, M. A., and Dikshit, A. K. (2011). Treatment of Landfill Leachate using Coagulation. *Proceedings from Second International Conference on Environmental Science and Technology IPCBEE,* 6, 119-122.

Shiskoswski, D. M., and Mavinic, D. S. (1998). Biological Treatment of a High Ammonia Leachate: Influence of External Carbon During Initial Startup, *Water Research,* 32, 2533-2541.

Silva, A. C., Dezotti, M., San't Anna Jr. G. L. (2004). Treatment and Detoxification of a Sanitary Landfill Leachates. *Chemosphere,* 55, 207-214.

Suthersan, S. S. (1999). *Pump and Treat Systems, Remediation Engineering: Design Concepts*. UK: CRC press.

Tam, N. F. Y. and Wong, Y. S. (1996). Effect of ammonia concentrations on growth of *Chlorella vulgaris* and Nitrogen Removal from Media. *Bioresource Technology*, 57, 45–50.

Tizaoui, C., Bouselmi, L., Mansouri, L., & Ghrabi, A. (2007). Landfill leachate treatment with ozone and ozone/hydrogen peroxide systems. *Journal of Hazardous Materials*, *140*(1-2), 316–324. https://doi.org/10.1016/j.jhazmat.2006.09.023.

Ujang, Z. (1997). *Case study 4: Hazardous waste management in Malaysia.* Retrieved 12 March 2015 from www.eolss.net/sampel-chapter/c09/e1-08-07.pdf.

Umar, M., Aziz, H. A., Yusoff, M. S. (2010). Variability of Parameters Involved in Leachate Pollution Index and Determination of LPI from Four Landfills in Malaysia. *Chemosphere,* 54, 1005-1010.

Vigneswaran, S. (2009). *Water and Wastewater Treatment Technologies.* DOLLS Publisher. UNESCO.

Wang, F., Smith, D. W., and Gamal El-Din, M. (2003). Application of Advanced Oxidation Methods for Landfill Leachate Treatment. *Journal of Environmental Engineering Science,* 2, 413-427.

Wang, Z., and Banks, C. J. (2007). Treatment of a High-strength Sulphate-rich Alkaline Leachate using an Anaerobic Filter. *Waste Management*, 27 (3), 359-66.

Wang, J. P., Chen, Y. Z., Ge, X. W., and Yu, H. Q. (2007). Optimization of Coagulation–flocculation Process for a Paper-recycling Wastewater Treatment using Response Surface Methodology. *Colloids and Surfaces A: Physicochemical and Engineering Aspects*, 302, 204–210.

Welander, U., Henrysson, T., Welander, T. (1998). Biological Nitrogen Removal from Municipal Landfill Leachate in a Pilot Scale Suspended Carrier Biofilm Process. *Water Research,* 32, 1564–1570.

Wiszniowski, J., Surmacz-Górska, J., Robert, D., Weber, J. V. (2007). The Effect of Landfill Leachate Composition on Organics and Nitrogen Removal in an Activated Sludge System with Bentonite Additive. *Journal of Environmental Management,* 85, 59-68.

Yang, Z., and Zhou, S. (2008). The Biological Treatment of Landfill Leachate using a Simultaneous Aerobic and Anaerobic (SAA) Bio-reactor System. *Chemosphere,* 72, 1751–1756.

Yilmaz, T., Apaydin, S., and Berktay, A. (2010). Coagulation-flocculation and Air Stripping as a Pretreatment of Young Landfill Leachate. *The Open Environmental Engineering Journal*, 3 (1), 42-48.

Young, J. C., and McCarty, P. L. (1969). The anaerobic filter for waste treatment. *Journal of the Water Pollution Control Federation,* 41, Suppl: R160+.

Yusoff, I., Alias, Y., Yusof, M., and Ashraf, M. A. (2013). Assessment of Pollutants Migration at Ampar Tenang Landfill Site, Selangor,

Malaysia. *Science Asia,* 39, 392–409. doi: 10.2306/scienceasia1513-1874.2013.39.392.

Zainol, N. A., Aziz, H. A., and Yusoff, M. S. (2011). The use of Polyaluminum Chloride for the treatment of Landfill Leachate via Coagulation and Flocculation processes. *Research Journal of Chemical Sciences,* 1(3), 34-39.

Zainol, N. A., Aziz, H. A., Yusoff, M. S. (2012). Characterization of Leachate from Kuala Sepetang and Kulim Landfills: A Comparative Study. *Energy and Environment Research,* 2 (2), 45-52.

Zhang, L., Lee, Y., Jahng, D. (2012). Ammonia Stripping for Enhanced Biomethanization of Piggery Wastewater. *Journal of Hazardous Materials,* (199-200), 36-42.

Zhang, F. (2015). *Reasons Why Landfill Leachate is Difficult to Treat?* Retrieved 2 February 2017 from https://www.linkedin.com/pulse/reasons-why-landfill-leachate-difficult-treat-fanny-zhang.

Zheng, T., Li, P., Wang, Q., Li, X., Ai, H., Gan, K. (2015). Pilot-scale Experiments on Brewery Wastewater Treatment and Sludge Reduction based on Food Chain Predation. *Desalination and Water Treatment*, 55 (5), 1142–1151.

ABOUT THE AUTHORS

Hasfalina Che Man is currently an academician at Department of Biological and Agricultural Engineering (DBAE), Faculty of Engineering, Universiti Putra Malaysia (UPM). She graduated with a Bachelor of Engineering degree in Chemical and Process Engineering, with special emphasis in Bioprocess, from the Universiti Kebangsaan Malaysia (UKM) in 1999. Soon after, she joined UPM and started her career as a tutor for a year. She was then given the opportunity to pursue PhD, and was conferred the doctoral degree in Civil Engineering and Geosciences from University of Newcastle upon Tyne, UK in 2005, specialising in Environmental Biotechnology, focusing on the applications of microorganisms for nutrient removal such as phosphorus and nitrogen from wastewater. In 2006, she began her career as a lecturer at DBAE, UPM. Since then, she has been involved in fundamental task of teaching and supervising. She had an experience to supervised more than 30 undergraduate students, 7 Masters and 3 doctoral candidates. She used to teach courses such as Agricultural Waste Management, Water Quality for Agriculture, Controlled Environment Structure, Engineering Mathematics, Agricultural Process Engineering and Environmental Biosystems. She has also served as the Final Year Project Coordinator, Internship Coordinator, Postgraduate Coordinator, and Head of Environmental Biosystem Lab.

Her primary contribution and main research interest is in the area of environmental technology, focusing on the adsorption materials from agricultural waste such as palm kernel shell, co-digestion of biowaste with biomass for hydrogen and methane production; and utilization of agricultural waste into valuable products such as palm oil mill effluent, rice straw or etc. More recently she also involved in mitigating climatic change involving identification of microorganisms generating methane in paddy field and methane emission control from livestock activities. She also involved on developing ammonia stripping treatment to treat real scheduled waste landfill leachate and decolorisation of palm oil mill effluent by physicochemical treatment. Other on-going research include bioconversion of wastes to useful biomaterials, in addition, effective microbs are being developed from local materials for improved composting process.

She has been appointed as a project leader and co-researcher for funding obtained from national awarding bodies such as MOHE (Ministry of Higher Education) and University Research Grant (RUGS, IPS, IPB), currently totalling up to RM 1.5 million. She also received an international TWAS research grant for monitoring groundwater quality due to livestock farming activities. Research findings have been translated into more than 45 papers and were published in national and international journals with impact factor/scopus/ISI. Research results were also presented as oral and posters at national and international conferences. She also had published a book and has been appointed as guest editor of selected journals such as

Pertanika, Acta Horticulture, Jurnal Teknologi and Agriculture & Agricultural Procedia. Recently, she has been actively involved in Malaysian Society of Agricultural Engineering (MSAE). She has been appointed as a Secretary General of MSAE for 4 years. Currently, she has been selected as a Vice President of MSAE. She is also a member of International Water Association (IWA), Institute of Engineers Malaysia (IEM), Board of Engineers Malaysia (BEM) and Associate Member for Institute of Chemical Engineers (ICheme).

M. Rashid is a professor of environmental engineering sciences, who is currently the Head of Air Resources Research Laboratory at the Malaysia-Japan International Institute of Technology, University of Technology Malaysia Kuala Lumpur, a higher learning institution that also provides research and consulting services in many disciplines. Dr Rashid was one of the founders of the environmental engineering program of the university and is now serving as one of its faculty members. A specialist in air resources engineering and management, Dr Rashid teaches and provides training and consultancy within his area of interest. He is a frequent speaker at seminars or workshops on the management of air resources and has served as a consultant to industries, consulting firms, and government agencies. He has written and published more than three hundred technical papers and reports in his area of expertise.

Dr Rashid is a registered EIA consultant, a member of the Malaysian Society for Engineering and Technology, Environmental Management & Research Association of Malaysia, and Malaysian Institute of Chemistry. A former member of the Environmental Task Force Group-Ministry of Science, Technology and Environment of Malaysia, Aerosol Society of UK, Air & Waste Management Association USA, and Association of Aerosol Research Germany.

About the Editor

Amimul Ahsan is an adjunct Associate Professor in the Department of Civil and Construction Engineering, Faculty of Science, Engineering and Technology, Swinburne University of Technology, Melbourne, Australia and Associate Professor in the Department of Civil Engineering, Uttara University, Dhaka, Bangladesh. Dr. Ahsan received a PhD in Civil Engineering from the University of Fukui, Japan and he has more than 11 years of research, teaching and industrial experiences. He is involved with a number of collaborative research projects in different countries, e.g., Western Sydney University, Australia; University of Fukui, Japan; University Putra Malaysia, Malaysia; University of Sharjah, UAE; Hamdan Bin Mohammed Smart University, UAE & Victoria University, Australia. He specializes in Water and Environmental Engineering, e.g., water and wastewater treatments, groundwater, water-air-soil pollution, solar desalination, rain-water and waste management. He has published extensively in Water and Environmental Engineering including 7 books, 10 book chapters and over 124 journal articles (mostly ISI/Scopus indexed). Sultan Selangor (Chancellor, UPM) has conferred him "Vice Chancellor Fellowship Award" (Science and Technology) in 2015. He has been serving in the editorial board of a number of refereed journals for a long time as an Editor-in-Chief (4), Editor/Associate Editor (6) & Editorial Board Member (11), e.g., *SpringerPlus* (ISI), *Membrane Water Treatment*

(ISI) & *Cogent Engineering* (Scopus). His *h* index in Scopus is 22. He was a former faculty in the Department of Civil Engineering and a former key researcher in the Institute of Advanced Technology (ITMA), University Putra Malaysia (UPM), Malaysia [2010-2017].

INDEX

C

D

E

F

G

H

I

J

L

M

N

S

T

U

V

W

Z